INCREDIBLE TECHNOLOGIES OF THE NEW WORLD ORDER: UFOS - TESLA - AREA 51

ABELARD PRODUCTIONS INC

COVER & INSIDE ART BY WESLEY CRUM

First Printing 1997
Special Limited Edition

©1997 by ABELARD PRODUCTIONS, INC.

All rights reserved, including the right to reproduce this book or portions thereof in any form whatsoever including (but not limited to) electronic, mechanical, photocopy, recording or otherwise. In all cases, we have tried our best to acknowledge the original source of the material utilized, except in cases where the parties have requested their true identities be concealed.

For foreign or other rights contact:
GLOBAL COMMUNICATIONS
Box 753, New Brunswick, NJ 08903

UFOS, TESLA, AND AREA 51 by Commander X

Contents

Introduction: Top Secret War 7
Chapter 1: Invaders in the Sky 10
Chapter 2: Nazi Technology 15
Chapter 3: The Alien Contract 21
Chapter 4: Inside Area 51 28
Chapter 5: Underground Bases 42
Chapter 6: Exposing Secrets 47
Chapter 7: Tesla's Death Ray 53
Chapter 8: The Russian Project 58
Chapter 9: HAARP 62
Chapter 10: Beam Weapons and the New World Order 68
Chapter 11: Killing You Softly 75
Chapter 12: Mind Arsenal 85
Chapter 13: Mind Control 91
Chapter 14: Alien Mind Control 106
Chapter 15: Far Reaches of Control 110
Chapter 16: UFO Base Discovered 128
Chapter 17: New World Order 132

UFOS, TESLA, AND AREA 51 by Commander X

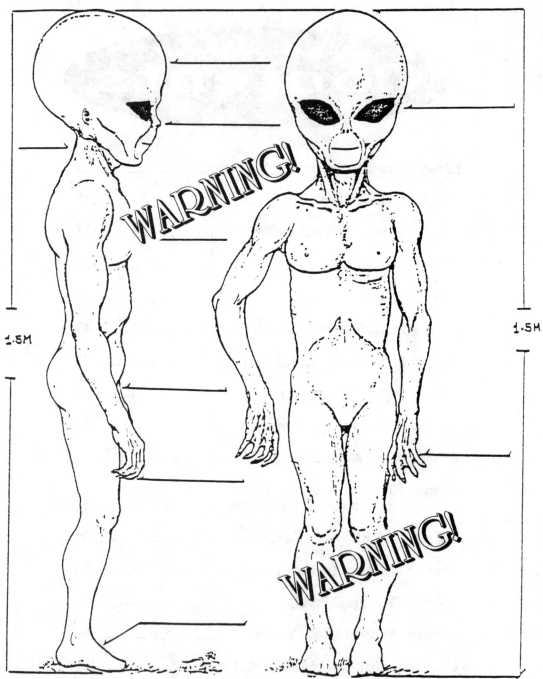

This sketch is based upon the testimony of hundreds of indiividuals who have personally had first hand encounters with the beings best known as the Grays, who for many years have been abducting humans and doing a variety of experiments on those who have been unwilling victims to what amounts to mass kidnappings for sinister purposes.

UFOS, TESLA, AND AREA 51 by Commander X

"A nation can survive its fools and even the ambitious. But it cannot survive treason from within. An enemy at the gates is less formidable, for he is known and he carries his banners openly. But the traitor moves among those within the gate freely, his sly whispers rustling through all the alleys, heard in the very halls of government itself. For the traitor appears not a traitor! He speaks in accents familiar to his victims and he wears their face and their garments and he appeals to the baseness that lies deep in the hearts of all men. He rots the soul of a nation. He works secretly and unknown in the night to undermine the pillars of a city. He infects the body politic so that it can no longer resist. A murderer is less to be feared."

--- Cicero, 42 B.C.

UFOS, TESLA, AND AREA 51 by Commander X

It is now more than speculation -- at least according to some "highly accredited" sources -- that various groups of extraterrestrials are coming to Earth, and many of them do not necessarily have our best interests at heart.

UFOS, TESLA, AND AREA 51 by Commander X

INTRODUCTION BY COMMANDER X
TOP SECRET WAR

There is currently a Top Secret war being conducted by the Secret Government and its allies, of human and alien races, against the people of planet Earth. I know that this is a startling proposition for many persons, but I am not known for flinching when revealing the truth. Although you will not read about it in the newspapers, or hear about it on the daily news, that does not make the conflict any less real or any less dangerous for the people of Earth.

Most of the humans who are participating in the battle for this world are not even known to you, eschewing fame for the sake of anonymity, and in most cases are not elected officials or even members of government. That does not mean that they do not wield enormous power, power that cannot even begin to be fathomed by those who have not delved into their incredible secrets.

These men, a small handful of unscrupulous individuals at the extreme summit of power on this planet, are currently consolidating their power and bringing to the fore weapons of control and destruction the likes of which have never been seen, at least on this planet, for the universe has a long and incredible history that has not even been scratched in contemporary accounts. These men work in close collaboration with the most deadly threat that mankind has ever faced, a threat that I have worked throughout my career to expose.

The main purpose of this book is a very simple and focused one. It is to familiarize you with some of the strategies of the Secret Government controllers and their alien masters, and to provide information about the new, incredibly high tech arsenal of weapons that is being used to manipulate and control us from behind the scenes.

It is to acquaint you with the purposes of the draconian regime of the New World Order and the terrible technologies which will be used to achieve the purposes of that group, unless these words are heeded and these men are stopped.

UFOS, TESLA, AND AREA 51 by Commander X

Reader: Do not make the mistake of thinking that the book that you hold in your hands is wild science fiction, or that it relies upon mere speculation when it comes to information about the Secret Government and their extraterrestrial allies. Those who are informed know that the truth about world conspiracy is far more amazing and sinister than what is broadcast and printed in the controlled media, which is in fact controlled by the Secret Government and its underlings. If anything, I have maintained a conservative approach to the information that I have presented in these pages, and I have held my tongue when I am not sure of the veracity of a report, verifying my sources and avoiding any guesswork.

This book is meant as a quick education in what is happening right now, at this moment, all over the world. It may be that portions of what is printed here may require that you expand your mind to accommodate the incredible truth that is only now beginning to be known. It is my belief that it is what you need to know at this time, so far as I have been able to determine it, and my hope is that you use this information well.

Time is short, and the future of the human race hangs in the balance in this cosmic battle which is mostly unguessed-at by the majority of persons on this planet. This is nothing less than a report on the covert war to control the Earth, and the means that will be used to effect that control, and there remains little time to ajust this fact. That some of the players in this Top Secret War are not even human is perhaps the hardest thing to swallow for someone who has not studied the subjects of UFO contact and, even more importantly, world conspiracy. But is it possible to ignore the testimony of hundreds and thousands of individuals who have had these experiences, and have risked their lives to expose the nature of this hideous collaboration? Reader, take your time and make your own judgement in this matter.

Use the information in this report well and in a timely fashion, for everything is at stake. We stand at the verge of a new world, if we can avert the New World Order.

UFOS, TESLA, AND AREA 51 by Commander X

Various Earth made disk are known to exist, some of which date back to World War II and German science.

UFOS, TESLA, AND AREA 51 by Commander X

CHAPTER ONE
INVADERS IN THE SKY

While the vast majority of the populations of the world are content to let their faith and their future reside in the hands of the forces that are beyond their control, and to let their attention be wasted on the electronic simulation of life that is television, there are a few of us who see ominous warning signs about the direction that this planet is heading in.

The UFOs which have been regularly observed in the skies of the world, with their numbers tremendously increasing over the last 45 years, are not the least sinister of these clues to what is going on behind the scenes.

Although not everything is known about the nature of these high technology flying craft, or about their alien (and human) pilots, enough is known to provide an understanding of their nature and purpose, and to make educated suppositions about the rest.

An all-too-brief rundown of what I have discovered follows.

SEEKING ANSWERS

I first began to understand the seriousness of the threat of the UFOs many years ago while working in the higher levels of American military intelligence. I saw the transition in American intelligence between World War II and the Cold War which followed, and I was not happy with many of the changes that I saw taking place behind the scenes. Incredibly, it became apparent to me that many of the changes that were being wrought were directly opposed to the purposes of American democracy and to the will of the American people.

As with many other military men who later delved into the nature of the Secret Government, men like top researchers Jason Bishop III, William Hamilton III,

UFOS, TESLA, AND AREA 51 by Commander X

"Branton," and Bill Cooper, I saw enough coverup and obfuscation going on of the phenomena of UFOs that it motivated me into serious research into this area and into asking questions that certain factions in military intelligence felt I should not be asking.

I was one of the first to realize that there were connections between the phenomena of UFOs and the gradually increasing factors of world control that I saw taking place internationally. To this day, many UFO researchers do not make this connection, out of fear or out of ignorance, I truly do not know.

At a certain point I woke up to what was going on behind the scenes on this planet, both in secretive meetings in ornate offices, and in underground bases around the world. Fortunately, I was able to pursue my researches, at least for a time, within the inner sanctums of the intelligence branch of the military. I was working underneath the noses of the conspirators, you might say.

During this time of fevered work and attempting to link the puzzle pieces of conspiracy, I also was able to make many valuable contacts within the hierarchies of military intelligence who believed the way that I do, and I am still in contact with many of those men and woman today. Many of them are uncredited due to the sensitivity of their positions.

At a certain point my alliance to the people of this planet rather than to the hidden controllers was discovered by those ranked higher in the chain of command, and I was forced to "retire," plainly at the risk of losing my life, as many of my compatriots warned me. It was thought that that would put an end to my meddling into matters that were secretive and not meant to be exposed to public view, but it did not turn out that way.

Since my retirement from military intelligence, I have continued to seek the truth in whatever ways that I have been able to do so, doing my best to pursue anonymity for myself, and without the ability to access intelligence agency computers and files, at least as readily as I was able to in the past.

In some ways, however, being a civilian has made it easier for me to pursue that quest for the answers, for I do not have to justify my movements or my inquiries to anyone.

UFOS, TESLA, AND AREA 51 by Commander X

My research eventually led me the answers that I sought, answers that were grim enough and incredible enough to cause me to wonder if steps toward the final climactic culmination of world control by alien forces might not have not proceeded a little farther down the road than I, or anyone, would have ever imagined.

There is, in fact, not a day that goes by that I do not wonder if things have not already proceeded too far down the road to One World Government and the completion of the plans of the Secret Government... and their inhuman masters. I try to keep negative and defeatist ideas like that in check, however, and to continue to work against the odds, as many, many dedicated people on the "right side" in this planetary battle do.

FIRST CONTACT

I have written of the early beginnings of flying saucer technology elsewhere, however much new information has recently come to light which I will share at this time. I will be touching only lightly on areas that I have covered in my writings elsewhere. It is vital that we understand as many threads of the cosmic treason that has taken place as possible, in order to trace the structure of control and deceit to its lair.

Although there has been sporadic contact between extraterrestrial races and human beings for many centuries and on many planets throughout this and other galaxies, contact that has been spoken of in the legends and mythologies of almost all ancient races of earth, this contact reached a culmination in the latter days of the last century. This was when German occult secret societies like the Thule and Vril societies initially contacted Grey aliens from Rigel who were allied to the extraterrestrial Orion group.

This early German society contact took place via a variety of ritual magic channelling and telepathic methods, as I said, at about the end of the last century. These modes of contact can easily be studied by reading some of the occult books that were prevalent in this period, works by men like Adolf Josef Lanz and Guido von List, although I would certainly not recommend that anyone attempt to duplicate the experiments that the members of these dark societies engaged in. One need only think about the terrible results of that first contact to realize what a dangerous proposition it can be.

UFOS, TESLA, AND AREA 51 by Commander X

Alien contact at least on the astral plane continued for several years (according to the annals of these secret groups, and the stories of the few members who defected from their ranks) and subsequently led to meetings between these secret society masters and the aliens on the physical plane, and an early exchange of information and perhaps technology of an advanced nature. There are rumors about the initiation of Adolf Hitler into these societies, although complete details are lacking and in many cases are contradictory.

The time was propitious for the Nazis, or so they thought. This alien exchange of information leapfrogged the capabilities of the Nazis in the air prior to World War II. This was the reason that the Nazis were able to forge ahead of the Allies with their research into flying saucer craft and exotic forms of energy production.

UFOS, TESLA, AND AREA 51 by Commander X

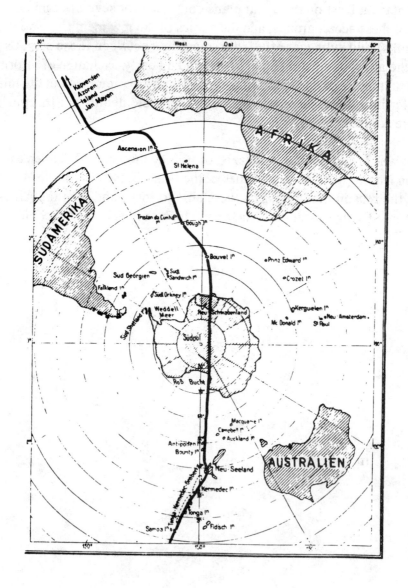

Before the last days of World War II the Nazi set up bases in the Antarctic from which they could carry on their efforts to control the world. This shows their expedition route.

CHAPTER TWO
NAZI TECHNOLOGY

As unbelievable as it may seem considering history as taught in colleges today, by 1944, in Germany, there was a fully workable flying saucer that had been created, developed by the Technische Adademie der Luftwaffe in their experimental centers. Other German craft were at various states of readiness, although it is rumored that there had even been successful missions to the Moon. I am unable to confirm or deny this, although there were sightings by astronomers of activity on the moon at this time that cannot have been natural in cause.

Other researchers also attest that these early extraterrestrial flights did take place, regardless if this flies in the face of conventional wisdom about when earthly space programs were initiated.

Configurations of German saucer craft included Victor Schauberger's free energy saucer craft and tachyon magneto-gravitic drive craft (termed the "Thule tachyonators" by its inventor), pioneered by Hans Kohler. All evidence points to the fact that, if the war had gone on much longer, such saucers as the 'Vril' series (named after the life energy termed thus by some German secret societies), which were 25 meter tank killers, and 300-foot in diameter 'Haunibu-4' interplanetary ships would have been activated and used in combat against the free world.

The leaders of the Allies cannot have guessed how close the free world came to being annihilated at this time, for there was also a crash course of atomic bomb development taking place in Germany at the time. Can you imagine the effect of atomic bombs carried by discoid craft?

Although I believe that researcher Vladimir Terziski is primarily speaking of capabilities that were developed by the Nazis during the first part of this century, as opposed to actual flight testing of these craft, there is the possibility that those capabilities were utilized in experimental flights. Terziski's account follows:

UFOS, TESLA, AND AREA 51 by Commander X

That Germans had in World War II saucers with no moving parts, that could go to the Moon in one hour, is not even the biggest secret... It is much more important that they managed to do that without using a single drop of traditional fuel for that purpose. The so-called free energy drives they were using were not perpetuum mobile drives, to use the dangerous word of the physics community, that were 'producing' energy out of nothing, but convertors of the practically inexhaustible for us gravitational energy of the Earth, Moon or other planetary bodies that these craft were flying by, into the electromagnetic energy necessary to propel the craft in space.

The well-known (in the field of alternative energy research) Hans Kohler convertor was utilized to convert the inexhaustible gravitational energy of a planetary body into the electromagnetic energy of the flight. The convertor was coupled to a Van de Graaf band generator and to a Marconi ball vortex dynamo in an electro-magneto-gravitic, or tachyon drive -- the so-called Thule and Andromeda tachyonators, that were mass produced on assembly lines at the AEG and Siemans plants in Germany in 1942-1945. These convertors were also used to propel their giant 5000 ton transport submarines or to produce energy in their underground bases around the world.

It is extremely fortunate that the Nazis, however, were not able to bring their experimental craft to the point where they were able to resist the Allies en masse (although the well-known Foo Fighters which were launched during the latter days of the war may have been German in origin, according to some intelligence sources). It is said that much information and even some working prototypes of the saucer craft were captured by the Americans, British, and Russians, after the close of the war, but like the stories of huge loads of Nazi gold captured at the same time, complete details on this are sketchy and are naturally closely guarded by those who do have the details.

These early German prototype saucers were hidden in underground facilities throughout Europe, including several locations in the Bavarian Mountains, and several pitched battles were engaged in by the Nazis and the Allies at the end of the war to take possession of these valuable prototypes and their plans for construction. This advanced information and technology was shipped to the United States, Great Britain, and Russia in the days immediately after the war. Other things were shipped, though...

UFOS, TESLA, AND AREA 51 by Commander X

One of the unfortunate things that took place after the war was that Nazi scientists were given a clean bill of health by certain highly-placed Nazi sympathizers in U.S. government and other elitist circles internationally, and these men were allowed to infiltrate the U.S. space program, and other important areas of U.S. government and the economy. It is unbelievable that men like Werner von Braun were allowed to suddenly profess a change in allegiance, but it suited the purpose of the men in control at that time to closely examine the beliefs of these Nazi scientists.

NAZIS IN ANTARCTICA

The end of the war did not entirely stop the Nazi thrust to power, unfortunately. Other factions within the Nazi hierarchy did not migrate to America.

At the end of World War II a Nazi base was established in the Antarctic by the elite S.S. group, and supplied by submarines carrying supplies and capable of travelling great distances underwater. Hitler and his wife Eva Braun, some say, did not actually commit suicide, but instead lived out their days beneath the ice of the South Pole, or in other accounts in the isolation of a secret retreat in South America.

This goes a long way toward explaining many of the saucer contacts involving German-speaking crews which have taken place in the intervening years, and which actually continue until the present. Especially in the early days of UFO contact by men like George Adamski, we hear of the occupants of these craft being tall, blonde, and nordic in features (and sometimes speaking German!), and this provides a clue to the origin of some of these discoid craft. It is quite possible that these contacts were with Germans, not Arcturans, and it is also probable that this secret Antarctic base exists to this day.

Certainly there have been rumors of a German Antarctic base, and more than one team of explorers in the area has "just disappeared" leaving no trace. Again, Vladimir Terziski fills in details about the German South Pole colony:

The Germans started seriously exploring the South Pole with huge carrier ships in 1937. The Schwabenland ship was sent to the lands of Queen Maud, south of South Africa, and the Germans promptly dropped their swastika flags from the planes and claimed the land, as big as the whole continent of Europe, as belonging to

UFOS, TESLA, AND AREA 51 by Commander X

the Third Reich. They gave it the name of Neu-Schwabenland. In 1942 a massive secret evacuation operation was undertaken by the Kriegs Marine to move people and materiel with submarines and ships to the secret underground base that was to become the last bastion of the Reich. Several hundred thousand concentration camp slave laborers, also scientists and young Hitler Ugend members, were evacuated with submarines and ships to the South Pole and extensive German colonies in South America, to continue the Nazi experiment at creating a pure racial society of Super Menschen. The rumors are that there is a vast underground city nowadays under the South Pole with a population of two million, that is called -- you've guessed correctly -- New Berlin. The major preoccupation of its inhabitants nowadays is human genetic engineering and space travel. Admiral Byrd is rumored to have secretly met with leaders of the German Antarctic colony in 1947, after his inglorious defeat, and to have concluded a condominium agreement with them about the peaceful coexistence of the German Nazi colony under the South Pole and the United States government, and for the exchange of advanced German technology for... American raw materials.

Further details about the Nazi South Polar base and German craft capable of leaving the earth's gravity are included in the book Man-Made UFOs 1944-1994, by Renato Vesco and David Hatcher Childress (Adventures Unlimited Press), which I would highly recommend due to its solid nuts and bolts treatment of the early days of discoid craft research:

Towards the end of the war [World War II], the Germans had developed interplanetary craft with no moving parts which were capable of going to the Moon or even Mars. Some researchers, in videos and articles, have claimed the Germans actually did this either at the end of the war or shortly afterward from their Antarctic Bases -- a fantastic claim!

Some sources claim that a "Last Battalion" of German soldiers escaped to Antarctica and South America via submarine in the last days of WWII. It is possible that the Germans may have high tech super cities in the remote jungles of South America as well.

A number of military historians, such as Col. Howard Buechner, author of Secrets of the Holy Lance and Hitler's Ashes, maintain that the Germans had already created bases in Queen Maud Land, opposite South Africa during the war.

UFOS, TESLA, AND AREA 51 by Commander X

Afterward, German U-Boats, in some reports as many as 100, took important scientists, aviators, and politicians to the final fortress of Nazi Germany. Two of these U-Boats surrendered in Argentina three months after the war. In 1947, the U.S. Navy invaded Antarctica, except for the German area known as New Schwabenland. On one flight [Admiral] Byrd's instruments went haywire and his entire squadron returned to their base via land navigation. It is often claimed that several jets from the four aircraft carriers which participated in Byrd's invasion were shot down by discoid craft. The Navy retreated and did not return until 1957. This time with the support of France, Britain, and Russia.

Other Nazi bases are presumed to have existed in isolated areas of South America, probably in mountain jungle locations, and possibly in the fjord area of southern Chile.

According to the book, Chronicle of Akakor, first published in German by the journalist Karl Brugger, a German battalion had taken refuge in an underground city on the borders of Brazil and Peru. Brugger, a German journalist who lived in Manaus, was assassinated in the Rio de Janeiro suburb of Ipanema in 1981.

UFOS, TESLA, AND AREA 51 by Commander X

Under a tight lid of security the U.S. military undertook to duplicate the technology of crashed UFOs. Much of this "space age" technology has been tested out in Nevada's desert in a spot that has become known in recent years as Area 51 (a place the government denies even exists).

CHAPTER THREE
THE ALIEN CONTRACT

As difficult as it may be for us to believe, accustomed as we are to thinking of extraterrestrials as being the stuff of science fiction, the Nazis had maintained a cool although continuous relationship with the aliens since their initial contact in the late 1800s. This contact is hinted at, although not admitted, in accounts of these societies written during this period. It would be many years before the aliens saw it as being in their best interest to expand their "diplomatic relations" with the people of earth.

The Americans, from our best information, established initial contact with the Grey extraterrestrials early in the 1950's, when two alien craft were accidentally disabled by radar in the vicinity of Aztec, New Mexico. In 1954 more formal auspices for meeting were established between the humans and aliens, when a secret meeting was arranged with then-President Eisenhower. This first meeting was accomplished by electronic translators which were provided by the aliens.

There is information about earlier contact with a humanoid extraterrestrial groups known as 'The Benevolent Ones,' as well as a reptilian race referred to as 'The Dracos,' that had taken place with the government of the U.S., but lacking in-depth information I will do no more than mention this fact and follow it up with a fuller account in the future.

What we do know for certain is that full-blown intergalactic interchange with the Greys began with President Eisenhower, although from interviews with people close to Eisenhower, it has come out that he was repulsed by the creatures and wisely always maintained a wariness about dealings with this extraterrestrial race. President Eisenhower, it is believed, although closely allied and later betrayed by factions of the Secret Government, did not entirely go along with their plans, at least in his private beliefs.

Although President Eisenhower rejected closer collaboration with the aliens at this time, diplomatic relations were started in order to prevent a conflict between earthlings and the extraterrestrials. It was an unavoidable compromise for the statesman.

In 1964, however, after Eisenhower was no longer president of the United States, a deal was cut with the aliens (including members of the oppressive Orion Group, which acts as the puppet masters for the Secret Government at this time) involving the transfer of information and technology, in exchange for certain freedoms they might exercise on Earth. This deal was engineered by the treasonous (and famous) MJ-12 group, and these freedoms included a number of terrible "deals with the Devil." We allowed the aliens access to vital resources, two of them being humans and cattle.

This is the genesis of cattle and human mutilations as sanctioned by the Secret Government. This deal was shepherded by the secret and treasonous MJ-12 group, existing outside of government, per se, and was an act of ultimate terrorism against the people of the world. The only way that this group of men, composed of scientists as well as men in the government, could have justified this deal even to themselves would have been through promises of power and technological wealth by the Greys, although it would be far from the last time that the interests of the human race would be betrayed.

Naturally, throughout this period, the government of the United States, infiltrated as it is by Secret Government loyalists, did not reveal anything to the public about our contact with alien beings, nor anything about advanced flying saucer research that had literally been going on for decades. This information has only started to come out through the dedicated efforts of civilian researchers, men and women who have not buckled under to the efforts to silence their communications.

Information was held Top Secret involving the exchange of many forms of technology that had been developed by the aliens. This included mind control technology and other forms of social control also offered by the aliens to make their own job easier, as will be brought out later in this book. Since this time the aliens (and in particular the Orion Group) have gradually and quite predictably extended their intrusion and power on the earth, including taking part in numerous joint-alien underground bases around the world, such as Area 51, in Nevada, and Dulce, in New Mexico, and certain entirely alien bases. Here incredible experimentation has taken place which will also be detailed later. Their UFO overflights and animal and human mutilations have continued since they are an integral portion of the deal that they made with secret factions of the U.S. government.

Yes, incredible technology for world control has been shared with the Secret Government... but not without a terrible price for humanity.

SECRET CRYPTO

Researcher Ed Komarek, Jr. provides important background details on contact with aliens and the possession of extraterrestrial saucer craft in his interview with a retired Air Force pilot:

This report describes an interview that I made on December 18, 1991 with the assistance of a close relative of the individual interviewed. The individual, the relative, and the location of this individual must be kept confidential to avoid official recriminations. This information is still classified above top secret and at this time has not been authorized by the Pentagon for release to the public. This man who I will call Jim for the purposes of this paper checked with his provost marshall at a nearby military base to see if he could talk to me. The provost marshall told him that much information was on the verge of being released and all was due to be out in five or six years. The marshall told him he would check with the Pentagon by computer and should have an answer in a couple of days. When Jim did not get back with his relative for an interview with me, I decided along with the relative to go see him immediately. When we caught up with Jim he said that the marshall had not got back with him yet.

He was willing to talk a little, though there was much he was not willing to tell us as he was getting concerned about the difficult position he was getting himself in. As he had already talked to the relative in more depth and I and the relative were later able to fill in some of the pieces that he would not tell me at this time. He had been threatened in the past that the military had places for people who divulged top secret information. He said that this information was classified at the highest level and that he had an above top secret crypto clearance and the need to know as he flew secret missions for the Air Force in Africa and elsewhere in the late 1940's and early 50's.

Jim said he was one of 59 individuals who were part of an elite group who flew these missions. He retired on a disability in the later 50s, but was kept on a retainer

UFOS, TESLA, AND AREA 51 by Commander X

until the mid-1980's.

His military records had been lost in a fire that destroyed the records of a large number of service men. This may account for why the provost marshall had not yet got back with him on this issue. Jim said that because of his position and the need to know he had access to the above Top Secret files that contained information on crashed saucers, pictures of flying saucers in flight, the flight characteristics of the saucers and pictures of the occupants of the saucers in military custody. In his secret missions he and the other men were expected and did come into contact with alien spacecraft. Jim also had an encounter with two craft and their occupants in the 1970's on a road near where he lives now. There is also continuing activity around him and his relatives in this local area.

At the time he was flying secret missions in the 50s there were 12 saucers in military custody. He had seen pictures of blond, blue-eyed aliens that looked like us in the files. He did not want to talk about the aliens or the pictures of them. I believe he said that Doolittle had taken the pictures, but I am not sure as he could have been talking about the saucer pictures. He was not personally familiar with the alien types collectively known as the Grays, but had seen drawings and perhaps pictures of them in a preserved state. I found that Jim was very well informed on these matters, and most of his information had come from his experiences in the military. He was not familiar with most of the literature available in the public sector, but had heard of the crash at Roswell which was one of the reasons that he was willing to talk about one of the crash film clips he had seen. I think he is seeing if he is going to get into trouble about this and may become more forthcoming in the future if nothing happens. I was concerned that after the marshall got back with him he might clam up, so I pestered him for as much information as I could get.

Jim talked with me about Ike's [President Eisenhower's] meeting with the aliens and said that Ike was given a tour of one of the ships. One craft in operating condition was given to the government at that time. He said that at that time when he was in the service the military could get the crafts to lift up, but had trouble getting forward motion. The military understood very little how the craft operated and least of all the powerplants.

He said his group had lost only one plane and that was probably because after an encounter with a spacecraft the plane's instruments had become perturbed by the ship's powerful magnetic fields and had come on to the runway from the south, when it was supposed to come in from the north. It blew up in a ball of fire at the end of the runway. Jim said he was never ordered to fire upon any of the saucers, yet said it was easy to shoot them down, even using Sidewinder missiles, at least that was what he had heard. The only time he knew that the military shot at the spacecraft and their occupants was when they were where they weren't supposed to be, as for instance over military installations. Jim had heard of one plane that was disintegrated by, most likely, ultrasound.

Jim says that he has seen several film clips of crashed spaceships and also of saucers skipping through the air like stones on top of the water. He would only talk to me about one five-six minute film clip of a crashed saucer inside a hangar. The main body of the craft was on the floor and the central unit or core was on a flatbed truck nearby. The craft on the floor had a 10 foot x 10 foot hole in the side with what looked like piping and conduit hanging out around the hole. He said there were lettering and symbols inside, but would not talk to me about the inside of the craft. He had told his relative that inside there were many control panels, what looked like star charts on the inside walls, and a large screen. He said to me that one room was still inaccessible to the military at the time the pictures were taken. The object was disk-shaped and he doesn't remember if there was a dome on it as the pictures were from one angle. The surface of the craft appeared dull grey. Jim was told that the craft was light enough to be picked up by one man and that the materials were very strong and pliable, but when pounded would return to their original shape. The material was extremely light.

The core which he believed to be the power plant was heavier and was still on the flatbed truck. He surmised that when the object struck the ground the heavier core punched through the wall of the craft, leaving the gaping hole. There were bodies on the floor of the hangar, but they were covered and they looked about five to five and one half feet long.

Jim says he has seen pictures of craft hovering over power lines and bodies of water, which he believes they use to fuel their craft. Also, he believes that they can get at underground water. He says that the sparks that his relative had seen around the outside of a UFO were sparks caused by dust from the atmosphere igniting from the intense temperatures in the plasmoid surrounding the craft.

His sighting in the mid-70s happened as he was driving down a local road near where he now lives. The two craft which were each wider than the road came head on and as they approached his car, slowed to a stop. The two craft hovered in the road ahead of his car. The craft had square windows and were shaped like a tire. The surface of the craft was as if one took golf balls and imbedded them in a surface, forming a hexagonal pattern. The craft had foot thick tubes around the circumference within which some kind of light energy seemed to be flowing. There were two beings of human form in each craft. They appeared to be identical with the same features and a suit of some flexible material that did not wrinkle when they moved their arms. It covered all the body except the face.

Let us hope there will be more information forthcoming from this case. It seems from some of the recent literature that some servicemen are being allowed to talk on this subject by their superiors. Perhaps this is a developing trend to allow information to come out on a gradual basis. I hope the Pentagon will allow this man to tell his story in full without recriminations so that other researchers may verify his name and confirm his story.

UFOS, TESLA, AND AREA 51 by Commander X

The secret government and the miliary learned much about the supernatural capabilities of the extraterrestrials by performing autopsies on these beings several of whom had died in crashes in the Southwest during the late 40s and early 50s.

UFOS, TESLA, AND AREA 51 by Commander X

CHAPTER FOUR
INSIDE AREA 51

By now almost everyone has heard about the secret military testing facility located in the Nevada desert and called Area 51, or sometimes Dreamland. It has become the fodder for many a national television show hinting about UFOs seen in the vicinity, but not daring to reveal the darker and more important truths about this Top Secret domain.

It has become so popular a spot for visiting UFO researchers that the highway that runs near it has now been dubbed the "Extraterrestrial Highway" and is visited by thousands every year. But there is another factual history of the place that is not often approached in television other media accounts.

This facility, Area 51, was first built and activated in 1954 (at approximately the same time as the beginning of the alien/U.S. compact). In these early days the CIA contracted with Lockheed to build the U-2 spyplane and to engage in other super-secret experimentation about which we can only speculate. Spy planes, however, were not the entirety of the research which has taken place.

Later, in 1962, testing of the SR-71 plane and the CIA's A-12 took place, along with a variety of anti-gravity research projects that had been developed from the researches of such men as the famous Albert Einstein and the much less well-known Nikola Tesla. Russian fighters also were tested at Area 51 (for reasons that may be speculated upon), while the F-117A first took flight in the area in about 1985.

It is vital to realize that these flights served only as a cover for far more advanced technical testing and experimentation that was going on in the Top Secret underground areas, including work that was connected to the Project Phoenix programs run by the military intelligence community, and the back-engineering of captured and traded-for alien saucer craft. This latter fact has been proved by the testimony of individuals who observed these craft both on the ground and in the air at the time, and by photographs taken during satellite overflights which were inadvertently supplied to researchers.

Almost the entirety of this super-secret compound is constructed underground, with numerous levels, laboratories, warehouses, and connecting elevators and roads, and a small army of military, scientific, maintenance, and intelligence services personnel manning the complex. Due to the large number of people that work at Area 51 to this day, there have been some people who have risked their lives to tell us about what goes on in this underground complex. It is from these cosmic patriots that the majority of our information is obtained, and to whom we are indebted.

If you were to ask the United States government you would be told that Area 51 doesn't actually exist, and that it is a hallucination of flying saucer buffs with overactive imaginations and too great an addiction to comic books. Also, if you consult FAA pilot charts and topographic maps issued by the U.S. government, Area 51 also simply doesn't exist.

According to these sources, all Area 51 is a remote tract of desert land called Groom Dry Lake where nothing takes place other than the meager activities of the indigenous wildlife. But that is not all that takes place at Area 51, by any means.

If it was, why would Area 51/Groom Lake be so closely guarded and electronically monitored? The area of Groom Lake is patrolled by the Wackenhut organization, a security group started by ex-members of government intelligence agencies and possessing more private dossiers on Americans than any other organization outside of the government itself. Wackenhut is known to have its fingers in many top secret pies internationally.

Approaching the boundaries of Area 51, not many people are willing to take their chances with the fences studded with solar-powered video cameras, motorized patrols, and ominous signs that read: "Use of Deadly Force Authorized." There are also numerous electronic sensors placed at intervals of a few feet that are used to detect unauthorized radio transmissions in the area, as well as other sensors that pick up the vibrations of cars that happen to trespass.

Stray, by accident or on purpose, near the top secret precincts of Area 51, and armed and hostile Wackenhut guards will arrest you and interrogate you about what you are doing on 'their' land, although stories are whispered among residents of the area of worse treatment that has been meted out to lone trespassers in the vicinity. It is verified that at least some persons claim to have experienced alien abduction in this area, and I have been present at several hypnotic regressions in which the stories were recounted about what took place after these individuals were taken underground.

On the other hand, it is incredible, but the government is telling the truth when it says that there is no actual government facility in the area. They base their statements on a mere technicality. In order to avoid the close scrutiny that would come about were Congressional oversight laws to be applied to what is going on, Area 51 is simply privately leased to a number of other bureaus and intelligence agencies such as the CIA and the Office of Naval Intelligence, as well as to private (although of the highest security) corporations like Lockheed.

CURRENT PROJECTS

In the past we have had verified a number of different secret projects taking place within the fenced confines of Area 51, with some individuals being willing to speak in detail on what is taking place.

According to the reports of people who work at this base, ships as large as "motherships" have been berthed and deployed from the location of Area 51, possibly even for extraterrestrial flights. UFO authority Sean Morton is one among many who have stated that flights to and from the Moon have taken place from this area. Films of these flights, as huge UFOs have landed or departed, crossing the mountains that surround Area 51, have circulated in the UFO research community.

Also, this has been the location for the creation of advanced, high technology electromagnetic detection and surveillance webs, and research into vortex and anomalous electromagnetic regions of the world which have been part and parcel of Project Phoenix, Montauk, and other secret projects of the Secret Government. Details will follow on some of these experiments.

This last research is part of the extensive thrust that the government has been involved in to utilize the researches of Tesla and Einstein, for such purposes as high tech weaponry, mind control and time and inter-dimensional travelling, although this will be treated more extensively later in this book.

UFOS, TESLA, AND AREA 51 by Commander X

Are we slipping into science fictional realms now, or is this actually happening in the real world? Reserve your judgement until documentation is offered.

What is going on in Area 51 at this moment? Tabloid television shows and mainstream magazine stories tease us with hints and allegations about what is taking place behind the fences of this Top Secret installation, but there are many aspects of this facility which still remain a mystery even to observers like myself who have attempted to ferret out the truth for years. Even top administrators at the secret base are treated on a "Need to Know" basis, not having the entirety of the scope of Area 51 projects revealed to them.

The mystery may in fact deepen in the years to come, due to the fact that the Air Force has recently closed off 4,000 acres of public lands surrounding Area 51, to prevent unauthorized onlookers from getting even a peek at this "Top Secret base that doesn't really exist."

Some of the more interesting things that have been happening at Area 51 recently that we know about are the testing of three huge 747-sized triangular-shaped craft, which some commentators have observed use technology back-engineered from alien -- that is, extraterrestrial -- craft.

Two gigantic hangars have also recently been erected at this location, both of them seven stories tall. What are these hangars to be used for? Some have said that these are not, in fact, hangars for extraterrestrial craft, but are in fact structures that enclose vortex entryways for dimensional travel.

Certainly there is no hint of advanced technology in the bland informational press releases from the government. We await more technicians involved in these projects to come forward and reveal the whole truth, although the personnel on this project are extremely tightly monitored and to be subjected to frequent lie detector tests to ascertain their loyalty to the Secret Government. Still, I would like to remind them that lie detectors can be beaten.

ADVANCED TECHNOLOGY

Some of the advanced, super-secret aircraft which I have personally confirmed as having flown out of Area 51 and associated locales include the high-speed

UFOS, TESLA, AND AREA 51 by Commander X

"Pulsar" craft, which, when heard, has an exhaust that "pulses." The Pulsar makes a very loud engine noise at about 1 to 2 Hz, and produces a linked "sausage-shaped" exhaust trail that is unmistakable when it is observed in the sky. It is said that the Pulsar can fly at speeds in excess of 5,000 miles per hour. The Pulsar also is sometimes seen to glow, looking like a single bright light, although sometimes also with a pulsating white or red light attached.

Another craft that is frequently observed in the area, and at numerous other Top Secret locations is a quiet aircraft, often observed being accompanied by Lockheed F-117A Stealths. This craft has been seen to appear and disappear in mid-air, causing some to speculate that it is intended to test Top Secret cloaking device capabilities.

Another craft, observed often and sometimes mistaken for an alien craft, is the diamond-shaped unmanned aircraft, which flies at high altitudes and supersonic speeds. This vehicle has been reliably identified as one that utilizes nuclear propulsion and a "gravity assist" to increase its speed and awesome maneuverability. Some of the banking maneuvers that this craft is capable of could not be withstood by a human pilot.

Although many secret aircraft that are flown from Area 51 utilize developments that take advantage of conventional jet engines, there are also craft that utilize breakthroughs brought about by scientific study of alien propulsion units, and information provided by alien collaborationist sources.

AREA 51 INSIDERS

Bob Lazar, a physicist allegedly employed at Area 51, is one of the few whistleblowers of a high technical proficiency to come forward with a coherent statement about what is going on behind the veil of secrecy at Area 51 and at nearby areas. He states that when he worked at S-4, located near Area 51, as an engineer he observed and worked upon several actual flying saucers of an extraterrestrial, not human origin, and that the government is currently working with Grey aliens at this facility. I have met Bob Lazar on a number of occasions in the past, and have spoken with him at length about the time he spent at Area 51, utilizing my own knowledge of the inner workings of secret government projects, and there is no reason to believe that he is not telling the truth.

UFOS, TESLA, AND AREA 51 by Commander X

Further background on Lazar is provided by Delve magazine, in an in-depth analysis of a radio interview done by Lazar:

1. Lazar's Purpose In Going On-Air:

Although Lazar's main stated purpose in appearing on the broadcast is to protect himself, caller Bill Cooper says one of Lazar's motives is anger -- previously expressed privately to Cooper -- that billions of dollars are needlessly wasted in the normal U.S. sectors by those without access to this alien technology. Lazar agrees.

Lazar says another reason for his appearing is to correct incorrect information he had heard on previous broadcasts.

Billy Goodman [the host of the broadcast] says Lazar's, Bill Cooper's, and John Lear's lives are on the line and that Lazar's best protection is the media, which is keeping Lazar alive.

Lazar hopes 'other people out there' working at S-4 [near Area 51] will loosen up, come forward, and join with him to present their information as one, as a group. He doesn't want to be the 'lone ranger.'

Congressional amnesty -- suggested by a caller -- would be nice, says Lazar, but merely offers freedom from prosecution. By coming forward, he concedes, the other S-4 workers have everything to lose and nothing to gain. But Huff [Gene Huff, a Las Vegas UFO investigator] says there has to be a moral guideline where national security has to hang in the balance and that the reporting of the existence of alien spaceships is where you draw the line.

Caller [and UFO researcher] Bill Cooper says there is 'a higher value' that those at S-4 should consider as a reason for coming forward to join Lazar.

But Lazar makes a point of distinguishing his whole view of the UFO situation from that of John Lear or Bill Cooper.

2. Verification of Lazar's Background:

Lazar says he worked at S-4 -- a 'restrained military environment' in Nevada -- from 12/88 to 4/89.

UFOS, TESLA, AND AREA 51 by Commander X

In response to callers who want independently to check up on Lazar, he says his 'colorful' background has already been checked by George Knapp of KLAS-TV, who had traveled to Los Alamos Laboratories and had spoken to former Lazar colleagues, who confirmed he really had worked there -- in spite of Los Alamos itself denying the fact of Lazar's employment with them. (Knapp's UFO broadcast displayed a page from a 1982 Los Alamos Labs internal phone book, listing Robert Lazar.) Caller Bill Cooper, who says he has talked to Lazar for over a year, says that by talking to persons at Lawrence Livermore Laboratory and at another facility, he and associates have confirmed Lazar's previous work in physics at the places Lazar specified. Lazar replied he was not aware of this checking done by Cooper.

Responding to an 'investigator' caller wanting Lazar to give him private info in order to check him out, Lazar says Lazar himself once took a correspondence course to be a private 'investigator.'

Lazar says he has 25 people each wanting independently to check him out, but he will not allow that.

Although Lazar admits he was paid by check, he refuses to discuss anything about the check stubs.

3. Lazar's Clearance Level:

He says neither REECO (Reynolds Electrical Engineering Co., Inc.) nor EG&G -- each a well-known U.S. Department of Energy Test Site contractor -- has people at S-4, that those persons' clearances are at most Q-Clearances, while his own clearance is "38 levels above Q-Clearance." The closest to S-4 that REECO or EG&G people physically get, he says, is Area 51.

4. Threats Made to Lazar/Lazar Shot At?

When Goodman refers to Lazar's telling him before the broadcast that Lazar had been shot at, Lazar says he doesn't want to talk about it.

Lazar says he was called to go back to work but officially refused because he didn't like the idea of returning to that isolated place in the desert where they could do what they wanted to him. However, he says his clearance has not been revoked.

Lazar says the executive powers at S-4 have run amok and there are no checks and balances. Congress, he says, has no knowledge of all this. Lazar says there are no lengths to which the military will not go to conceal this information.

5. Mind Control Suspicions:

He contacted hypnotherapist Layne Keck of Seranus Clinical Hypnosis in Las Vegas because there were a couple of days where Lazar remembered only going out in a plane and coming back, but nothing in between. He suspected mind control had been performed on him.

Under hypnosis, Lazar recalled intense drilling, threatening actions taken against him, and his drinking of "pine," which his hypnotherapist said was similar to the "Orion Method" of regimented hypnosis used by the military.

Per Huff, Lazar was given drugs and hypnosis by his employers -- not so he would forget what he was working on, but so -- by their imprinting his subconscious -- he would be afraid to talk.

Per Huff, after first telling Lazar his phone was tapped, the military later threatened him because, having monitored his phone, they knew he was planning to release information about the alien craft. Huff says the military were amazed that the drugs and hypnosis had not worked.

Although he is afraid, Lazar hopes they won't come after him now since he has already talked.

6. Location of the Saucers:

Lazar says the nine saucers are not at the supersecret Area 51 (Groom Lake) of the U.S. government's Nevada Test Site, but at S-4 -- 10 miles south of Area 51. The disks are only at this one place. However, to go to work, Lazar flew (by plane) to Groom Lake, waited a short time at a cafe, then got on a bus with blacked-out windows.

UFOS, TESLA, AND AREA 51 by Commander X

7. Bad Aliens Killed S-4 Workers:

Lazar says the aliens are not benevolent. Some humans were killed in a conflict after a U.S. military intelligence power play, after which point a previously ongoing information exchange ended.

Huff says Lazar earlier told him -- and Lazar agrees he had said this -- that this exchange of information occurred between the aliens and human scientists/human security personnel.

But although the aliens had allowed themselves to be under constant human guard, the aliens had insisted there be no bullets in the guns worn by the security people.

The security people ignored that demand, and they all died from head wounds -- which left no evidence of how they had died.

The aliens even killed the scientists they were teaching...

Responding to a caller, Lazar says talking about aliens is NOT a touchy subject for him. Aliens exist, he says.

8. Time of Saucer Test Flights:

Lazar says the experimental UFO flights at S-4 are flown either by remote control or by human pilots -- not the aliens.

10. Origin of the Saucers at S-4:

The alien vehicles Lazar saw up close in hangars come from "another world... the fourth planet out from Zeta Reticulum, a binary system."

11. Description of the Saucers:

He says he touched one of the nine UFOs and even stood in its doorway.

He says the saucers mostly appeared "like new" -- one of them looking like Billy Meier's saucer. Lazar says one of the saucers "looked like it was hit with some sort of a projectile. It had a large hole in the bottom and a large hole in the top with the metal bent out like some sort of, you know, large caliber 4 or 5 inch had gone through it."

The inside of one craft appears made of wax and then cooled off, all like a cast or mold of one thing with no rough edges.

Inside were small chairs, one or one and one-half feet, as though made for little kids.

12: Nine S-4 Extraterrestrial -- Not Earthly?

The nine disks he saw up close -- including the one he worked on -- were not earthly and were definitely extraterrestrial.

Lazar can't say whether or not the other disks he saw at a distance during testing are manmade or extraterrestrial.

13. Physics of the Saucers at S-4:

Lazar says the vehicle attaches itself to a distorted portion of space-time and returns with the distortion. It's a new physics. The vehicles brought a space-time warp with them.

Lazar says the propulsion technology should properly remain classified, since everything there is looked at from a weapons point of view. A lot is directly applicable to weapons systems, and he has no intention of releasing it, he says.

But he says the craft uses gravity as a lens and the power source is an anti-matter reactor.

Two Modes of Travel for Saucers at S-4:

1) When traveling around the surface of the planet, the vehicles balance on a gravity wave or ride a wave like a cork on the ocean. In this mode they are unstable and are affected by the weather.

UFOS, TESLA, AND AREA 51 by Commander X

2) For space travel, they use gravity generators. But if they fly around the surface of Earth using this mode they may flip over, a phenomenon Lazar says has frequently been observed in past sightings in the published UFO literature.

Two Gravities: A & B.

Gravity A works on an atomic scale whose interaction is small and has to do with fuel -- the alien Element 115 used for the disks [a previously unknown element utilized in the alien propulsion systems].

Gravity B works on a macro scale. The gravitational field is out of phase with matter and is like a wave generator. It is longitudinal generation -- not spherical, as a caller suggested.

14. Project Aurora:

"Aurora," says Lazar, distinguishing it from any UFOs, is the replacement for the SR-71 plane. It uses a three mile runway and makes a sound like continuous explosions. It has speeds up to Mach 10.

15. Few Abductees?

More persons claim to be abductees than have actually been abducted, suggests Lazar.

16. No Blue Diamond Entry Way for Saucers:

Notwithstanding UFO watchers congregating at Blue Diamond, Lazar says there is only a gravity anomaly around Blue Diamond, not an entryway for saucers into our universe.

17. Other Physics Comments by Lazar:

Time Travel: Gravity affects time. Moving forward in time is "a breeze." All you need to do is get close to a gravitational field. Moving back in time might also be possible.

UFOS, TESLA, AND AREA 51 by Commander X

Some physicists today, says Lazar, invoke superstring theories [of physics] to simply add another dimension to the universe whenever they can't explain something.

Another anonymous source, connected with Congress and with a high security clearance, has visited Area 51 but was only allowed to visit the top levels of the multi-level facility. According to this source,

"This is not part of the official program of the U.S. government. Although aircraft are being tested and flown at government ranges, I think this is some sort of intelligence operation, or there could be foreign money involved... it's expensive and is immune to the oversight process. This defrauds the American government and people. You go to jail for that."

SURVEILLANCE

One of the amazing surveillance techniques which is used at Area 51 and at other locations are small anti-gravitic "UFOs" or "orbs" that some researchers have reported following them when they are on or near these restricted areas. I myself have observed these small surveillance units at a distance near Area 51 and at other locations, although I would not recommend attempting to come into close contact with them for a variety of reasons which should be obvious.

One researcher, J.S., in Santa Ana, California, has done a good deal of research on this aspect of New World Order technology, although certain of his conclusions are not shared by me. As far as these small robot UFOs, there is the possibility that they may some day may move off of the Top Secret reservations and be utilized to control the masses in the city and countryside. This possibility has been spoken of in government reports on what they call "Non-lethal weaponry."

J.S. reports,

I have taped pictures and sounds of the very visible objects in the sky which I have titled 'alien light balls.' I believe the sounds are communication just as we have Morse code but of course I can't know their meanings. I have recorded about 63 video tapes so far. I'm sure that the Marfa [Texas] Mystery Lights are controlled by computer-like creations but I feel there are other areas besides Marfa, Texas that have the same phenomena. It is not inconceivable that the round lights are cameras,

UFOS, TESLA, AND AREA 51 by Commander X

eyes, whatever. I have given these sightings the name of 'Robotic Controlled Moving Eyes' and they are fascinating. They are very fast and can be sighted very low to the ground with tremendous speed for gaining altitude. It is possible that they are a form of atomic nuclear energy also. I understand that NASA has been researching this for a very long time and may have had contact with extra-terrestrial beings. It seems that the metal or metal products of these sightings are not visible during daylight hours. Perhaps they are all around during the daytime but are not capable of being seen.

I understand there are bases (also underground) in New Mexico, Nevada and Costilla, Colorado where it is rumored there are actual remains of the 'grey-beings' but of course such installations would be under super secret security.

I have visited Tonopah, Nevada a couple of times but of course I could not enter the base. In driving around the area at night I experienced those same hovering balls of light with the beeping sounds that are audible on my tapes. further on, some of the white, cylindrical lights came very close to my car, blinking lights and approximately five feet from me. I photographed as much as I could and got some pretty good pictures and sounds. I also saw two round objects, about three feet in diameter and they had red and green lights around the top. In scrutinizing some of my tapes I sometimes think features can be deciphered but I am not sure of that. I do know that for a very long time, and my tapes will verify this, strange objects and sounds are over the area I live.

According to numerous abductees who have literally been kidnapped by aliens, in many instances they have seemingly been taken to underground bunkers or bases where they have been probed, examined or sometimes even impregnated.

UFOS, TESLA, AND AREA 51 by Commander X

CHAPTER FIVE
UNDERGROUND BASES

Although a complete survey of underground bases is not possible in this book due to the length of treatment that would be required, another example of these Top Secret bases is deemed "The Anthill" by those in the know. This underground base is located in the Rosamond Valley portion of Antelope Valley, in California, in the area of the Tehachapi Mountains. There exists a secret Northrop facility, consisting of numerous underground levels and extensive connecting tunnels and motorized vehicles for easy travel between these secret underground locations.

This is an area of routine sightings of UFOs of various sorts, along with frequent overflies of the mysterious black helicopters, which have often been linked to cattle mutilations and other mysterious events. There are many workers at the Anthill who live in the area, and they can be occasionally coaxed into talking about what they have seen in the rooms and corridors that honeycomb the area underground.

William Hamilton, in his Alien Magic (Global Communications, Box 753, New Brunswick, New Jersey, 08903) relates what he was told by "Joe," who works in the Anthill facility:

Joe was in construction and held clearances for working in military operations areas. He said that he had worked on an underground tunnel project below Haystack Butte on the eastern boundary of Edwards near the NASA Rocket Test Site. he also claimed that he saw orbs roving around these tunnels. He painted numbers inside a box locate on a stripe that ran horizontally midway along the tunnel walls. I asked him how far this tunnel ran below the earth's surface. He said that he and fellow employees used to count as the elevator descended to the tunnel level. From the count and from the elevator speed, I estimated that the tunnel must have been around 3,000 feet deep.

UFOS, TESLA, AND AREA 51 by Commander X

One time he saw a door open to a room in one of these tunnels and he could see a very tall alien standing next to two men in white lab coats. He thought his alien was all of nine feet tall. he claims that he saw two grey aliens inside a hangar at China Lake one day when he went back inside the hangar after finishing his work to retrieve a tool that he had left behind...

One day Joe told me about two old school buddies he had run across. They both held jobs in underground facilities and had worked at the Anthill, he said. They would work underground for two weeks at a stint. They lived in condos when working underground. These condos were also built into the underground facility. The government even picked up the tab on one guy's alimony. One was known as a computer genius. He said that he had seen both grey and reptilian life forms in various underground facilities. One of the underground projects was Project Startalk. The work involved lasers.

The informer, a guy named Paul, said that he worked in a big underground building (350 feet across). Project Startalk utilized a powerful laser which strikes a mirror and is sent into space. The laser is modulated with a signal and acts as a beacon to bring in UFOs. Apparently, the beacon is directed at friendly forces from other star systems. He also worked at the Douglas facility near LLano. He once saw a saucer land and go into an underground hangar. Inside the underground building is a huge computer complex. The workers wear white clothing and white socks (no shoes). The computers us an alien symbolic language. Manuals indicate codes that can be entered. There is large lexon plastic screen in this complex that displays various star systems and galaxies. A wax pencil is used to indicate targets for the laser. The technology used is so far advanced that it is beyond known engineering technology. The laser is also capable of interdimensional communication. Security workers accompany all workers, even to the bathroom. Phones are tapped, even workers' home phones.

ORB MONITORING

Hamilton also describes one particularly telling encounter that took place in this area:

Ray and Nancy worked at the Northrop B-2 assembly facility in Palmdale [California]. Ray is a Native American. Ray was an aircraft inspector and worked the swing shift. One June night he decided to take a midnight ride with Nancy up to the cut in the Tehachapi Mountains. This cut appears as an inclined whitish mark on the side of the foothills. It actually marks the site of a road that winds up around the mountains. On the backside is an entrance to an artificial plateau that had been blasted out of the rocks. Ray parked his truck on this plateau. They got out to look at the stars and the city lights of Lancaster and Palmdale in the valley far below. While looking at the stars, Nancy noticed that some of them were moving around and brought it to Ray's attention. Ray got his flashlight out of the truck and started signaling the lights. At some point, Nancy noticed a bright basketball-sized orb hovering just above a nearby knoll. They both walked closer to get a better look at the orb. Ray thought it had just risen out of some invisible opening in the ground. It seemed to be flashing and sparkling. Some sort of line dangled from its underside.

It rose a little higher and Nancy tried to speak to it, having an intuitive feeling that some intelligence had guided the orb to that plateau for their benefit. As they watched this strange phenomenon around one o'clock in the morning, they next observed the dawn light over the far distant eastern hills. Something had just snatched four hours of time out of their lives. The orb was gone. They were terrified and drove quickly home. The next day, they felt a vibration going through the apartment. When they went outdoors, they saw two orbs hovering above their apartment. This scared them badly.

I took Ray and Nancy to a local hypnotist and she regressed Ray. Nancy refused to be regressed, expressing fear over what she might discover about those four hours of missing time. Ray was an excellent subject. When in trance, with little prompting, he fell backwards nearly to the floor before we caught him. The regression brought out some amazing revelations. Ray and Nancy had been abducted and taken underground!

Under hypnosis he kept mentioning the Kern River to the north. "There is an area near the Tehachapi Mountains called the Kern River Project. The upper river is being used by the government for hydroelectric power to power the underground facility at Tehachapi Ranch (actually, the Tejon Ranch). The mountain next to the power facility is being hollowed out... there is mud all around and it's so obvious, but apparently people aren't looking. All the power is being used for the Ranch, which is the sit of underground 'skunk works' where highly technical aircraft, spacecraft and all kinds of stuff are being dealt with.

UFOS, TESLA, AND AREA 51 by Commander X

"It is a huge underground base, probably close to the size of one under 29 Palms Marine Base. It has huge hangars and very large elevators as well as technical laboratories. There is a whole city under there, large passageways... the whole valley is full of tunnels. You can drive from one end to the other underground. You can drive from Palmdale, site of the Northrop, Lockheed and 'black project' areas, to California City, all underground. There are tunnels all the way to George AFB. Aliens apparently have access... they've been seen all over the place. "The government lets them do whatever they want. They're probing the human brain, trying to find out weaknesses and learning whatever they want. They're probing the human brain, trying to find out weaknesses and learning how to control us. They dissect humans... can't describe the dissections because they are not humane. Really morbid. The government knows it... they just turn their heads. Some people in the government want to stop this but they don't know how to stop it."

Ray was disconcerted that Greys had Nancy strapped to a table in this facility. He could see instruments all around. During hypnosis he would freeze up when recalling this scene. He yelled and became very emotional. He was convinced they would rape her and violate her, yet he was helpless to prevent it. He and Nancy also felt they had contact with a benevolent race of aliens who had observed their capture. After a few months, Ray and Nancy announced to me that they wanted little more to do with reporting further events. They felt exposed and monitored and feared retribution if they continued talking.

From several sources we have been informed of a super secret group called MJ-12, composed of high ranking government officials, military personnel and scientists who have worked alongside of various alien groups since the first UFO crashes in the southwestern United States.

UFOS, TESLA, AND AREA 51 by Commander X

CHAPTER SIX
EXPOSING SECRETS

There are far more high tech wonders (and horrors) that are currently being utilized by the New World Order in order to defend their planetay property and chattel, and to create the kind of 1984-like future that they desire, for their intention is to stop at nothing less than total control of the people of earth.

For more background on the conspiracy and its alien and human collaborators, we again consult William F. Hamilton III, one of the very few researchers who is able to piece together the whole fantastic picture. In a privately-circulated article titled, "The Exo-Biological Intervention (EBI) Hypothesis," Hamilton pulls together divergent accounts regarding the conspiracy and speculates about the ultimate purpose of the aliens and the Secret Government, reaching the only conclusions that can be reached. His account follows:

Historical:

Various groups of human, humanoid and non-human races came to earth in the distant past and colonized our planet. There were, no doubt, squabbles among the different groups and races. One or more of these groups bred the species of modern homo sapiens into existence for purposes of their own. Tis explains why there is no missing link found between primates and humans. Probably due to interspecies and intergroup conflicts, we were freed of our original bondage to our creators.

Bear in mind that on a spiritual level of existence some of us may have been our own creators, but on a physical biological level, the earth human is an artifact of genetic engineering and genetic manipulation. There are non-terrestrial humans who have near-perfect genetic breeding and whose life spans exceed our own by at least a factor of ten. They have perfect sight, hearing, and form. They are telepathic and intelligent. We are a poorer model of this same type human.

UFOS, TESLA, AND AREA 51 by Commander X

Other species, perhaps the Grey humanoid, has bred humans and cross-breeds for its own purposes. RH-factor in blood seems to have been introduced by the Greys.

There is a good possibility that Draconian reptoids are another superior species who came to our world in the distant past. Our biblical records mention the ANGELS (Nordic humans) and the SERPENT race (Dracs) as ancient enemies.

These early space visitors not only generated us biologically, but influenced our social and cultural patterns which we have inherited today. Those institutions created directly by ET races were intended to control our growth and development. These institutions are mainly in the form of government and religion.

The Beginning of the Modern Era:

The spacecrafts that crashed at Roswell and Aztec, New Mexico and recovered by secret military and scientific teams was the known beginning of the modern era of ET presence in our world. In fact, they have been with us to one degree or another since ancient times residing in permanent installations underearth and underwater. The first contacts of the modern era may have taken place as long ago as 1933.

The ANGELS returned to contact man on a one-to-one basis when the first physical contact was made with George Adamski on November 20th, 1952. Prior contacts made by them were made in private. The contact with Orthon was staged as a public event with witnesses.

The first public revelation of humanoid abduction came after 1961 when the Hill abduction occurred. The demons and jinns of old had also returned.

The Means:

The two principles used in protecting the activities of ETs on Earth are SECRECY and DECEPTION. If the human race have been the victims of manipulation, then that manipulation was carried on under the cloaks of secrecy and deception. It was a game that men in government and religion (including secret occult societies) were to learn well from our hidden masters and controllers. In addition they added the elements of RIDICULE and DENIAL.

UFOS, TESLA, AND AREA 51 by Commander X

The Cover-Up:

The cover-up was initiated soon after the Roswell, NM crash. We wanted to know 1) who they were 2) why they were here, and 30 how their technology worked. The cover-up became a matter of NATIONAL SECURITY (a blanket word covering secrecy and deception). The cover-up involves secret organizations within our government such as MJ-12, PI-40, MAJI, Delta, the Jason Scholars, and known intelligence organizations such as Naval Intelligence, Air Force Office of Special Investigation, the Defense Investigative Service, the CIA, NSA, and more.

It involves THINK TANKS such as RAND, the Ford Foundation, the Aspen Institute, and the Brookings Institute. It involves corporations such as Bechtel, GE, ITT, Amoco, Northrop, Lockheed, and many others. It involves SECRET SOCIETIES who may be hidden bosses of the orchestrated events -- the Illuminati, Masons, Knights of Malta, etc. The individual players are too numerous to list. The whole of this conspiracy forms an interlocking nexus. The goal is said to be a ONE WORLD GOVERNMENT.

The Underground Nation:

The RAND symposium held in 1959 on Deep Underground Construction indicates that plans were hatched during the fifties to build underground bases, laboratories, and city-
complexes linked by a stupendous network of tunnels to preserve and protect the ongoing secret interests of the secret societies. These secret societies made a pact with alien entities in order to further motives of domination.

The underground complexes are not confined to the U.S.A. alone. A large underground complex operated by the U.S. exists at Pine Gap, near Alice Springs, Australia.

Alien Abduction Scenarios:

Unlike the friendly contacts made by some of the human and humanoid aliens, there are those abductors who treat the earth human much as we treat a high form of animal life. The humans are taken without being given a choice, examined with specimens taken, sometimes inseminating females, tagged or implanted with monitors, and released with some amnesia. The purpose of these abductions seems to serve the motives of the aliens and serves no beneficial purpose for the abductee. However, not all alien contacts conform to this scenario and there do appear to be alien beings who do want to assist us on some limited scale. There is still much to be learned about the aliens.

Planetary and Environmental Catastrophe:

It appears that the secret societies among us have become aware of coming planetary eco-catastrophe and the possibility of an earth polar shift in the near future. Surveying the earth from space, satellites and shuttles reveal extensive damage to our ecosphere. Our planet is wobbling on its axis and its magnetic field is decaying. Ozone depletion and the greenhouse effect are rapidly endangering life on our planet. Alternatives, which include 1) direct handling of the atmospheric problems 2) taking shelter in underground domains, and 3) escape to other planetary bodies in the solar system have been devised in secret. However, there is a possible Alternative 4 which mostly depends on a completely different idea on how to save the earth.

We must also bear in mind the possibility of conflict or war over the earth between different clashing groups. If the earth is a prize among all the different power groups, then it might be saved. We must face the insane possibility that a doomsday weapon may be put to use (Cobalt bombs or anti-matter bombs).

The Future -- Despair or Hope? (A Call for Unity)

If all races and all groups persist in playing the old covert games, then we can only despair over our future. But if all are willing to transcend their allegiance to a group or life form and form a greater society based on Love - Understanding - Integrity - Openness - and Freedom for all, then we can visualize a flicker of hope.

Love is the axis around which life turns and the road the wheel turns on is the road of knowledge and the destination is perfection and truth - The Truth Revealed. We must free ourselves from the need for governmental and religious institutions for all can govern their own life, keep their own counsel, and all can experience religious freedom and truth for themselves. It is imperative in order to survive that everyone grant total freedom and self-determinism to everyone else.

UFOS, TESLA, AND AREA 51 by Commander X

This is the only way out for all of us or we will continue to enslave and entrap one another. Only when we can have understanding among all the people of earth and those not of this earth will peace prevail.

A New World Vision:

In order to build a new world from the ruins of the old world we need to evolve as individuals. The spiritually free person, free of psychological and emotional problems, advanced in intellect, intuition, and wisdom can bring a new harmony to the world. The new person will be healthy and free of lust, greed, and envy. The new person will be able to form new social units of survival that may not resemble the nuclear family of the pat age -- it will be a new kind of tribe for the welfare of the many. Men and women will enjoy equal status. Children will be shared. A new society of creative workers will evolve. Mankind will build his cities and industries underground and the surface of the earth will revert to natural growth. Our technology will be an eco-tech based on the fundamental forces of the universe. When the people of earth live and love in harmony as one then will be explore the distant stars as well as the depths of the spirit worlds and advance far beyond primitive homo sapiens, the children of the stars.

We need a ONE WORLD COMMUNITY not a ONE WORLD GOVERNMENT!

UFOS, TESLA, AND AREA 51 by Commander X

Seen here in his New York lab, many thought that Tesla had gone to his grave without giving away secrets that would enable the military to put the finishing touches on his "Death Beam," a potentially destructive device.

TESLA, AT 78, BARES NEW 'DEATH-BEAM'

Invention Powerful Enough to Destroy 10,000 Planes 250 Miles Away, He Asserts.

DEFENSIVE WEAPON ONLY

Scientist, in Interview, Tells of Apparatus That He Says Will Kill Without Trace.

Nikola Tesla, father of modern methods of generation and distribution of electrical energy, who was 78 years old yesterday, announced a new invention, or inventions, which he said, he considered the most important of the 700 made by him so far.

He has perfected a method and apparatus, Dr. Tesla said yesterday in an interview at the Hotel New Yorker, which will send concentrated beams of particles through the free air, of such tremendous energy that they will bring down a fleet of 10,000 enemy airplanes at a distance of 250 miles from a defending nation's border and will cause armies of millions to drop dead in their tracks.

NOTED INVENTOR 78.
Nikola Tesla.

CHAPTER SEVEN
TESLA'S DEATH RAY

Nikola Tesla is one of the most interesting and enigmatic figures ever to live in this world, and if anyone among the human race deserves the title of being a Star Being, it was he.

Tesla lived at about the same time as the far-better known inventor Thomas Edison, who many suggest was far over-rated in comparison with his less well known contemporary, and in fact the two men disliked each other and engaged in a heated competition to develop new scientific technologies -- all to the benefit of earth, at least at the time.

For his part, Tesla wanted to help the people of earth with his discoveries, one of which was a means for transmitting electricity through the medium of the earth itself, broadcasting electricity through the ground to distant locations. It was a means of delivering electrical energy that would have been entirely free for the consumer, and it would have been impossible for any individual or cartel to place a price tag on it. Unfortunately for all of us, his lofty plans were never to achieve fruition, thanks to the covert activities of the Secret Government.

The controllers would have never accepted a free energy source for mankind, for that would have broken up their extremely lucrative monopoly on energy. In 1905 a patent was issued to Tesla describing how to "Transmit Electrical Energy Through the Natural Mediums" but the famous and wealthy financier J. P. Morgan (who had financed both Tesla and Edison, monopolizing electrical technology worldwide), severed communications and funding for Tesla. At the same time he launched an intensive anti-Tesla media smear campaign, labelling him as a demented crackpot and blackmailing others who were interested in funding his researches.

Tesla was to die penniless and almost completely forgotten by his contemporaries, while his unpublished research was to be confiscated by American intelligence agencies at the time of his death. So much for his ideas of providing free energy to the people.

TESLA'S DEATH RAY

Many of the advanced and secret technologies currently secretly in place (as well as being developed) by the Secret Government owe their inspiration to research originally done by the genius Tesla.

One of these technologies is the death ray. Again, this is not science fiction that I am talking about, but a technology that currently exists. Tesla had done work on broadcasting immense power through the air as far back as the turn of this century. The New York Times, in fact, ran an article on precisely this Tesla technology in 1915. The article follows:

Nikola Tesla, the inventor, has filed patent applications on the essential parts of a machine the possibilities of which test a layman's imagination and promise a parallel of Thor's shooting thunderbolts from the sky to punish those who have angered the gods... Suffice it to say that the invention will go through space with a speed of 300 miles a second, a manless ship without propelling engine or wings, sent by electricity to any desired point on the globe on its errand of destruction, if destruction its manipulator wishes to effect.

"It is not a time," said Dr. Tesla yesterday, "to go into the details of this thing. It is founded upon a principle that means great things in peace; it can be used for great things in war. But I repeat, this is no time to talk of such things. It is perfectly practicable to transmit electrical energy without wires and produce destructive effects at a distance. I have already constructed a wireless transmitter which makes this possible, and have described it in my technical publications, among which I refer to my patent number 1,119,732 recently granted.

"With a transmitter of this kind we are enabled to project electrical energy in any amount to any distance and apply it for innumerable purposes, both in war and peace. Through the universal adoption of this system, ideal conditions for the maintenance of law and order will be realized, for then the energy necessary to the enforcement of right and justice will be normally productive, yet potential, and in any moment available, for attack and defense. The power transmitted need not be necessarily destructive, for, if distance is made to depend upon it, its withdrawal or supply will bring about the same results as those now accomplished by force of arms."

Tesla sporadically continued work on fine-tuning his death ray, and eventually, upon his death, his secrets were handed over to the United States government, classified Top Secret, and given restricted access except to those with highest security clearances.

Again, the New York Times commented, in 1940:

> Nikola Tesla, one of the truly great inventors who celebrated his eighty-fourth birthday on July 10, tells the writer that he stands ready to divulge to the United States Government the secret of his "teleforce," with which, he said, airplane motors would be melted at a distance of 250 miles, so that an invisible Chinese Wall of Defense would be built around the country...
>
> This "teleforce," he said, is based on an entirely new principle of physics that "no one has ever dreamed about," different from the principle embodied in his inventions relating to the transmission of electrical power from a distance, for which he has received a number of basic patents.
>
> This new type of force, Mr. Tesla said, would operate through a beam one hundred-millionth of a square centimeter in diameter, and could be generated from a special plant that would cost no more that $2,000,000 and would take only about three months to construct.
>
> The beam, he states, involves four new inventions, two of which already have been tested. One of these is a method and apparatus for producing rays "and other manifestations of energy" in free air, eliminating the necessity for a high vacuum; a second is a method and process for producing "very great electrical force;" the third is a method for amplifying this force, and the fourth is a new method for producing "a tremendous electrical repelling force." This would be the projector, or gun, of the system.

UFOS, TESLA, AND AREA 51 by Commander X

The voltage for propelling the beam to its objective, according to the inventor, will attain a potential of 50,000,000 volts. With this enormous voltage, he said, microscopic electrical particles of matter will be catapulted on their mission of defensive destruction. He has been working on this invention, he added, for many years and has recently made a number of improvements in it.

What greater proof than this admission by the New York Times would be necessary to prove that the United States has far greater technological capability in weaponry than its agents are willing to admit?

One of the technologies developed by Tesla has in recent years received much attention (as well as covert development). Tesla referred to this development as a radio-frequency (RF) oscillator, and one of the applications for this was his death ray, although the entire scope of Tesla's research in this area are still not fully understood, even by scientists.

The basic idea is to broadcast power into the atmosphere, as well as to focus it for varied results. It is known that both the United States and Russia have borne technological fruit from Tesla's first studies in this area, and most of the technology which has been referred to as "Star Wars" capability has, in fact, been based on the researches of Tesla. Scientist Dr. Marc Seifer, has made the remark that: "Great support is lent to the hypothesis that Tesla's work and papers were systematically hidden from public view in order to protect the trail of this top secret work, which today is known as Star Wars."

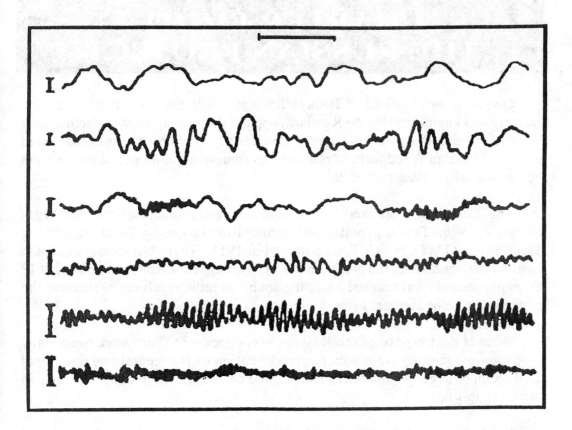

Here is a brain transmitter read-out showing that an individual mind can function quite differently under a variety of stimulas and control methodology.

CHAPTER EIGHT
THE RUSSIAN PROJECT

One of the applications of Tesla technology which has actually come on line recently was engineered by the Russians, apparently in an attempt to decimate the Western world through electromagnetic broadcasting. The project has been referred to as the "Russian Woodpecker," although its somewhat humorous name does not betray its deadly serious capabilities.

The "Russian Woodpecker" is a series of electromagnetic broadcasts that began on July 4, 1976. This apparently was accomplished by using Tesla Magnifying Transmitters (TMT), which Tesla patented in 1914. These broadcasts disrupted communications all over the planet, in the 3 to 30 megahertz bands, and pulsed at 10 times per second. This caused a tapping sound on radio receivers, which was the source of the "Woodpecker" name.

What is the purpose of the Russian Woodpecker? There have been many speculations, ranging from weather modification or for mental and emotional destruction of target populations. The purpose of the Russian Woodpecker may in fact be multi-fold.

We do know that the northwestern United States was continuously irradiated with Extremely Low Frequency waves that were set at a level designed to match biological frequencies, in other words to duplicate life energies of the target population. Even NBC admitted, as of July 18, 1981, that the purpose of the Russian Woodpecker was to influence human behavior.

One of the main Tesla magnifying transmitter sites discontinued its broadcasting in 1986 due to the explosion of the Chernobyl nuclear reactor, outside Kiev, Ukraine, which was its main power source. Some observers have even suggested that Chernobyl may have been sabotaged by the Western powers, in order to disrupt the Woodpecker broadcasting.

UFOS, TESLA, AND AREA 51 by Commander X

In 1978 Specula magazine provided further information on the Russian Woodpecker, saying: "an electromagnetic signal of certain frequencies can be transmitted through the Earth which, when introduced into the Earth at multiples of 30 degrees, will form standing waves in the Earth, and in certain cases induces coherence to the wave in the molten core of the Earth. One can induce earthquakes at a distant point, and severe atmospheric disturbances over the target area."

Again, this is not just wild speculation that I am offering. When Tesla was testing his broadcast power at the beginning of this century it set off lightning storms that resulted in power outages and hundreds of forest fires.

Need more convincing? According to Dr. Andrew Michrowski, Ph.D., at one time Technologies Specialist with the Canadian Department of State, and the President of the Planetary Association for Clean Energy,

Since October 1976 the Union of Soviet Socialist Republics has been emitting extremely low frequency signals from a number of Tesla-type transmitters. Their frequencies correspond to brain-wave rhythms of either the depressed or the irritable states of humans --and scientifically tenable tests have shown that the U.S.S.R. signals do lock-in human brain-wave signals.

The U.S.S.R. signals have been assessed by the Environmental Protection Agency ...to be psychoactive (i.e. liable to produce psychological response and vulnerability in humans). The same agency has noted that the U.S.S.R. ELF signals can be absorbed and re-radiated by 60-Hertz power transmission lines and even be magnified by water-pipe grids.

The Soviets are on the verge of a breakthrough into a new weapons technology that will make missiles and bombers obsolete. It would allow them to destroy up to five American cities a day just by sending out radio pulses. They could induce panic or illness into whole nations.

One side effect of the Russian Woodpecker signals which is not often noted is that of ozone depletion in the atmosphere, which was discovered in 1977, one year after the Russian Woodpecker went on line. There is a great deal of evidence to prove that these kinds of broadcasts disrupt the Earth's ozone layer.

UFOS, TESLA, AND AREA 51 by Commander X

According to an anonymous researcher, the Russian Woodpecker had its equivalent technologically in America. This was the "American Buzzsaw." In an interview, the researcher reported,

The "American Buzzsaw" is the U.S. equivalent of the Russian Woodpecker signal. This is a psycho-active signal which is designed to interface with the mind of the human being by way of the mind-brain connection. The government has been quite active, designing systems that are carried by helicopter, systems that sit on the ground like the buzzsaw transmitters, etc.

I first picked up the buzzsaw signal about 1990. I called the FCC and asked them if they knew what it was. They said it was the American version of the over-the-horizon radar. I said, "you mean like the Russian woodpecker signal?" They said "yes." So they admitted that it had the same purpose.

This signal is broadcasted on multiple carriers. The carriers hop from one frequency to another, anywhere in the range from 4MHz to 30MHz. It is never on the ham band or the international broadcast band. It is only on the allocated frequencies that the government shares with other communication services.

With the switching of these frequencies, they are creating what is known as a Levinson Transform, named after Norman Levinson, who generated the mathematics for frequency-time transformations. This is highly significant, since the human brain, body and mind work on time dependent pulsations and frequencies which are time encoded. You have this unusual pulse modulated signal hopping around from frequency to frequency to frequency. They have a multiple transmitter. The signals only come in phase at a targeted site.

This signal, we believe, has three modes. The first mode is the search-mode. This is where they transmit a signal and then a psychically sensitive human being picks up the signal and sends something back. They can in this way identify these people and where they are. The second mode is the general transmission of a psychic interrupt function which interrupts psychic activity. That's all it does. It has a tendency to lower the overall psychic awareness of the population. A good psychic can easily screen this mode out of their consciousness. The third mode is where they target an individual, and we have known six people who have been targeted. They can lock on to the resonance of the synthetic material in your clothing and target you from that. And then there is the horrific state of the art in electromagnetic broadcasting....

TOP SECRET   MAG CHANNELS ONLY

DEFENSE INTELLIGENCE AGENCY

PSYCHOTRONIC WARFARE: SPIRITUAL ACCESS

Prepared by U.S Army
Medical Intelligence Office
DST-03447/82/018

TOP SECRET   MAG CHANNELS ONLY

Documented proof that the military is researching metaphysical concepts and probing ways to use such belief systems to their advantage in matters of mind control.

CHAPTER NINE
HAARP

When researchers first came into contact with materials relating to HAARP (the High Atmospheric Auroral Research Project), they realized that Tesla technology had finally come to pass, but that it was a bitter fruit that the researches of this gentle scientist bore.

The main patents for HAARP were issued in the 1980s to Bernard J. Eastlund, for "Method and Apparatus for Altering a Region in the Earth's Atmosphere, Ionosphere, and/or Magnetosphere" Eastlund was employed by ARCO, a subsidiary of Altantic Richfield.

A GIGANTIC RAY GUN

Although you will be hard-put to find out much about the capabilities of HAARP by reviewing the New York Times or the Washington Post, what this project amounts to is a huge electromagnetic ray gun aimed at the upper atmosphere of the earth, and capable of being aimed and focused with great precision, in other words an electromagnetic weapon of tremendous capability.

HAARP is a transmitter that could be likened to a giant microwave oven with its beams capable of being focused at virtually any point on earth. Although the scientists who work on this project refer to it as an "ionospheric heater," HAARP has many more capabilities than simply heating selected portions of the upper atmosphere.

Simply put, HAARP is a system of electronic broadcasting that is currently being run by the military in the United States. One should be tipped off by the fact that this is a project of the military, not a civilian one as press releases would have you believe. While there are a number of capabilities to which HAARP can be addressed, the one that is the most important is never talked about... anywhere. That incredible capability is mind control.

One of the ominous things about this fact is that America is bound by treaty to bow to the control of the authority of the United Nations. What this means is that HAARP, ultimately, is under the control of the New World Order's mouthpiece, the U.N. and the use of this gigantic electromagnetic projector is completely not beholden to the people of the United States.

MENTAL DISRUPTION

Don't immediately recoil and come to the conclusion that Commander X has gone off the deep end with the information that I have just provided. I can substantiate everything that I say, in far greater detail than is possible here. For instance, consider the following report, penned by Dr. Nick Begich and Jeane Manning in Nexus magazine. According to the authors:

...US Air Force documents revealed that a system had been developed for manipulating and disrupting human mental processes through pulsed radio-frequency radiation (the stuff of HAARP) over large geographical areas.

The most telling material about this technology came from writings of Zbigniew Brzezinski (former National Security Advisor to US President Carter) and J.F. MacDonald (Science Advisor to US President Johnson and a Professor of Geophysics at UCLA), as they wrote about the use of power-beaming transmitters for geophysical and environmental warfare. The documents showed how these effects might be caused, and the negative effects on human health and thinking.

The mental-disruption possibilities for HAARP are the most disturbing... For example, one of the papers describing this use was from the International Red Cross in Geneva. It even gave the frequency ranges where these effects could occur -- the same ranges which HAARP is capable of broadcasting.

Begich and Manning quote Brzezinski, who wrote the following while a professor at Columbia University more than 25 years ago:

UFOS, TESLA, AND AREA 51 by Commander X

"Political strategists are tempted to exploit research on the brain and human behavior. Geophysicist Gordon J.F. MacDonald -- specialist in problems of warfare -- says accurately-timed, artificially-excited electronic strokes 'could lead to a patter of oscillations that produce relatively high power levels over certain regions of the Earth... In this way, one could develop a system that would seriously impair the brain performance of very large populations in selected regions over an extended period...'

"No matter how deeply disturbing the thought of using the environment to manipulate behavior for national advantages, to some the technology permitting such use will very probably develop within the next few decades."

They also quote MacDonald, who wrote a variety of papers describing this mind control technology. According to MacDonald:

"The key to geophysical warfare is the identification of environmental instabilities to which the addition of a small amount of energy would release vastly greater amounts of energy."

Begich and Manning wonder in their article:

"While yesterday's geophysicists predicted today's advances, are HAARP program managers delivering on the vision? The geophysicists recognized that adding energy to the environmental soup could have large effects. However, humankind has already added substantial amounts of electromagnetic energy into our environment without understanding what might constitute critical mass."

Ponder the significance of the following, issued by the U.S. Air Force:

"The potential applications of artificial electromagnet fields are wide-ranging and can be used in many military or quasi-military situations... Some of these potential uses include dealing with terrorist groups, crowd control, controlling breaches of security at military installations, and antipersonnel techniques in tactical warfare.

"In all of these cases the EM (electromagnetic) systems would be used to produce mild to severe physiological disruption or perceptual distortion or disorientation. In addition, the ability of individuals to function could be degraded to such a point that they would be combat-ineffective. Another advantage of electromagnetic systems is that they can provide coverage over large areas with a single system. They are silent, and countermeasures to them may be difficult to develop."

Naturally it is only whispered amongst groups of top military strategists in the Pentagon, that one of the capabilities of HAARP is to disrupt the populace within the United States. For that matter, even the simple exposure of the fact that HAARP is a weapons system, and not a scientific experimental project would be likely to set off a great wave of grass roots protest in the United States. That is why the government has been so careful to maintain a declassified status on the program, and to present it as simple research on radio capabilities, and an attempt to study the upper atmosphere of the earth.

HAARP CAPABILITIES

According to The Apocalypse Chronicles,

The truth is, the HAARP system represents a veritable Pandora's Box of engineering and electronic warfare capabilities. It is not a single weapon; it is an exponential leap into a basic technology that encompasses many applications that include weapons. In its full power implementation [scheduled to come on line in 1998, although some researchers believe that HAARP has been tested at full power already] the HAARP system can accomplish the following:

-- Completely eliminate and/or cripple all military or commercial communications systems throughout the entire world.

-- Intercept and decipher any and all communications systems that have been rendered inactive.

-- Actually control the weather over an area the size of a country, a state, or an extended region.

UFOS, TESLA, AND AREA 51 by Commander X

-- Implement a focused death ray technology that can discriminately destroy targets at great distances.

-- Focus a low intensity invisible beam with great precision impacting individuals that can accelerate cancer and other lethal conditions -- without the victim ever realizing they are being murdered.

-- Put an entire population center to sleep -- or cause them to become so emotionally agitated that they become violent and self-destructive.

-- Beam radio transmissions directly into individual's minds so that they believe they are hearing the voice of God -- or any other personality the people behind the transmission choose to masquerade as...

One of the ways in which **HAARP** can be used as an offensive weapon is in selective weather control of targeted areas. As early as 1958, a spokesman for the White House made the statement that the Defense Department was 'studying ways to manipulate changes of earth and sky, and so affect the weather.' Experiments were later made in cloud seeding to cause rain, but at the same time possibilities for doing this with electronics (specifically the kind of electronics developed by Tesla) were first studied. This paralleled the experiments with extremely low frequency (ELF) broadcasting transmitters, culminating with the incredible capabilities of **HAARP**.

HAARP CHRONOLOGY

For the researcher interested in tracking the progression of this technology, here is a brief chronology of significant events regarding electromagnetic warfare capabilities of the New World Order:

1886-1888: Nikola Tesla formulates Alternating Current and means for its transmission. At the time Thomas Edison insisted that DC current was the future for the transmission of electricity, although since that time he was proven wrong, given that alternating current is far more widely used today.

1900: Tesla applies for patent to "Transmit Electrical Energy Through the Natural Mediums," i.e. through air, water, and earth. This was the genesis for the technology which would be employed in future electromagnetic broadcasting technologies, including that of America's HAARP.

1938: In this year a scientist put for the proposition of illuminating the night by electron gyrotron heating broadcast from a transmitter. Again, this technology would later be turned to less humane uses by the military industrial complex.

1940: Tesla announces that he has invented a full-blown "death ray." This information was turned over to the U.S. government at his death, or before it.

1958: The announcement that the U.S. military is studying means of weather manipulation is made at this time. One of the considerations of the military was that this could be done by electromagnetics, and they had far more wide-ranging electronic prospects than weather control on the drawing boards.

1960: Beginning about this time the Earth experienced many catastrophic weather changes and shifts which, at the time, were unaccounted for. Now we have a partial reason why the weather suddenly went crazy: electromagnetic broadcasting and experimentation.

1974: Experiments on electromagnetic broadcasting which evolved into HAARP technology took place in this period, at Plattesville, Colorado; Arecibo, Puerto Rico, and Armidale, New South Wales.

1975: Research is released showing that ELF broadcasts alter human blood chemistry.

1975: Congress urges the military that any weather modification experimentation and programs be overseen by a civilian agency. The military ignores the suggestion.

1975: This was the period when the Russian Woodpecker ELF broadcasting system came on line, firing electromagnetic energy across the horizon at the U.S., with the energy specifically pulsed to duplicate human brainwave rhythms.

1976: In this year scientists demonstrated that nerve cells were affected and could be damaged by the broadcasting of ELF waves. One of the manners in which this technology was utilized was by the Russians beaming electronics at the American Embassy in Moscow, causing damage and illness to the workers there. Little protest was ever raised in this matter.

1980: Bernard J. Eastlund, who did much of the preliminary work and obtained the patents for **HAARP**, in this year obtained a patent stipulated as, "Method and Apparatus for Altering a Region in the Earth's Atmosphere, Ionosphere and/or Magnetosphere."

1980's: During these years the U.S. created a network of Ground Wave Emergency Network (GWEN) towers, broadcasting Very Low Frequency waves, allegedly for defense.

1995: Congress budgets $10 million for **HAARP**, specifically to be applied to "nuclear counterproliferation."

1994-1996: First stage testing of the **HAARP** broadcasting arrays -- at least that is what the published accounts say. Other researchers have suggested that HAARP came fully on line at this time, engaging in a number of projects and targeting various areas of the globe with its radiations.

1998: In this year HAARP is scheduled to be fully on-line, according to published accounts.

Throughout my career I have been called an alarmist and worse, and I will no doubt again be called an alarmist for my assessment of the HAARP project in Alaska. What is overlooked by my critics is that an alarm is precisely what is required at this juncture in our history, when truly awesome weapons of mind control and destruction are being brought on-line, with no assurance that they will not be used against We The People.

Let the ostriches continue with their heads buried deeply in the sand. Let them continue that way until their tailfeathers are fried by **HAARP**.

Is the New World Order preparing to launch an all out war on humankind?

UFOS, TESLA, AND AREA 51 by Commander X

CHAPTER TEN
BEAM WEAPONS AND
THE NEW WORLD ORDER

In the past twenty years we have seen the development of a vast infrastructure of electromagnetic and other exotic forms of weaponry, being deployed both by the military establishments of the United States and Russia. These form the crux of a rapidly-expanding capability of total control on the planet, control which will probably not be utilized for the good of mankind.

If it was in the interests of our nation that these kinds of weapons be constructed, weapons like HAARP and the Russian Woodpecker, why have we not been told anything about them, except through covert sources?

Richard Boylan, Ph.D.'s investigations of advanced New World Order weapons enlighten us on many weapons and research applications that you will not hear spoken of in the popular press:

The immense sprawling complex of Sandia National Laboratories [in New Mexico] with its test ranges that extend south and east of Department of Energy headquarters for more than one hundred square miles, takes up most of what is labeled "Kirtland Air Force Base." SNL has project buildings every half mile in every direction out to the horizon. Activities identified by the signs include mostly weapons applications research in nuclear, nuclear transport, magnetic, solar, electromagnetic pulse, laser and particle beam energy. At the Solar Power Tower laboratory SNL boasts a heliostate (a solar collector-concentrator device) that can burn through a 1-inch thick hard metal plating in 26 seconds.

But the piece de resistance of Star Wars weapons research application was Project Aries, the Advanced Research Electromagnetic Pulse (EMP) Simulator Site, where a two block-long device was built for the Defense Nuclear Agency by Edgerton, Germhausen & Greer (EG&G).

UFOS, TESLA, AND AREA 51 by Commander X

The EG&G Corporation is involved (along with Wackenhut Corporation) in security for Nevada's Area 51 and S-4, U.S. supersecret aerospace vehicle test centers (personal communication with security officers at EG&G and at Wakenhut 03/14/92), in the "black budget" (Black Budget is a term used to describe secret funding of special military or other covert projects not identified in the Congressional budget) weapons operations like Project Aries, and in maintenance of various U.S. Government nuclear facilities. The EMP weapon consists of a 1 1/2 block-long barrel horizontally supported on a wooden (nonconductive) trestle 25 feet high, connected to a two-story tower building, connected in turn to an immense electrical apparatus with huge arms and massive connecting cables that looks like a gigantic Van de Graaf generator. The long-rumored electromagnetic pulse weapon has been spotted at last! Using fusion power and engineered to 100 trillion volt bursts, it could arguably overpower even the most exotic extraterrestrial UFO technology.

FURTHER WEAPONS APPLICATIONS

Boylan reveals further exotic weapons applications that he has discovered in the course of his research:

Farther west in New Mexico, I investigated the National Radio Astronomy Observatory (NRAO) whose Very Large Array consisted of 27 huge 82-foot-wide "receiving" dishes, on support towers that move on railroad tracks. Each was as tall as a 12-story building, and capable of being grouped into many different configurations. The dishes I saw were arrayed to form an inverted "T", with each arm stretching for one mile; the longest arm pointed due north.

The stated purpose of this facility was to collect "weak radio waves from celestial sources". To an outsider, this meant mapping the heavens by locating stars and energized gas fields in space that emitted electromagnetic (EM) radiation in the radio frequency. However, as with other sites on this tour, NRAO was not an average observatory. Parked adjacent to headquarters was an Army truck and two ambulances with NRAO insignias (The handful of astronomers who worked here must have had terrible occupational accident records!)

Further insight was provided by NASA Ames Research Center spokesperson Dr. Jill Tarter during a presentation at the University of California, in Davis,

UFOS, TESLA, AND AREA 51 by Commander X

California on November 26, 1991. Dr. Tarter revealed that on October 12, 1992, the United States would announce it was turning on its radio telescopes to listen for intelligence transmissions from space. This announcement was times to coincide with the 500th anniversary of Columbus' "discovery" of the New World.

Wonderful, except that this is disinformation. Actually, the U.S. government had been funding and conducting the Search for Extraterrestrial Intelligence (SETI) for years. One could speculate that the purpose of this high-profile government announcement was to create an excuse to reveal ET communications, and finally an open admission by the government of extraterrestrials.

What was not speculation was the evidence I observed regarding the use of powerful signals into space; these dishes were directed just above the low north horizon.

A clue came while I was having dinner. Seated next to four NRAO astronomers, I overhead one complain about trying to get time on a radio dish to do his research. While I was at NROA all 27 dishes were pointed away from the main part of the sky, and directed instead toward a target above the low-north horizon. A photo of NRAO displayed at their headquarters also showed all 27 dishes pointed at the same low-north horizon angle. Why was there such a persistent focus on one area of sky when there was such competition for time on the dishes? Another clue was the desolate location of NRAO on the San Augustin Plains, a quiet region deliberately chosen for its remoteness from city radio stations and other EM radiation.

I obtained the physical evidence as I left NRAO. About two miles from the Very Large Array, with both my FM and CB radios on, the most excruciatingly intense screeching sounds came out of one of the radios while simultaneously blanketing both. My CB meter was farther into the red than I have ever seen it go before. The howling screeched for several minutes. I could not believe my ears. Wouldn't such powerful EM signals interfere with listening to delicate radio waves from space? I turned off the radios until I reached Pietown, 21 miles to the northwest, where, upon turning both radios on, I again heard the deafening screech. Mercifully, after about two minutes it stopped; both radios again functioned perfectly, sending and receiving. I have not heard that noise since.

UFOS, TESLA, AND AREA 51 by Commander X

It occurred to me then that NRAO is not only receiving signals from space, but also sending them!

Dr. Boylan reaches some remarkable conclusions in regards to the development of advanced beam weapons in the service of the New World Order:

Since 1992, a shadow government has continued its secret development of space weapons in response to intelligence species visiting Earth from other star systems. In 1993, at Kirtland Air Force Base, USAF's Phillips Laboratory rushed to complete a contract for the development of a trillion-watt electromagnetic pulse weapon. Also in '93, the Russians offered to share with the United States their advances in deployment of a plasmoid space weapon. A plasmoid is an extremely high-energy concentrated force field that can tear anything apart. This incredible system directs two very powerful energy beams: one consists of electromagnetic energy in the microwave range; the other consists of intense laser energy. These beams converge in space at a designated target -- something like a miniature hydrogen bomb inside the Superdome.

Obviously these weapons are not in response to the now passe Soviet ballistic missile threat. What then is the explanation? The time has come for this administration to come clean with the American people about what Star Wars (AKA Ballistic Missile Defense Organization) really is and what the government really knows about UFOs and their occupants.

UFOS, TESLA, AND AREA 51 by Commander X

Those who have been "inside" the New World Order insist that it is possible to alter and control brainwaves at considerable distances, thus enabling the "controllers" to make individuals act and behave in a manner which is unlike their normal behavior (even to commit crimes).

CHAPTER ELEVEN
KILLING YOU SOFTLY

Unbeknownst to the vast majority of Americans, the New World Order has been quietly preparing its arsenal of incredible weapons for the time, the inevitable time, when they will attempt to overthrow the sovereignty of the United States. An abundance of evidence shows this to be the case, and this purpose is well documented in the communications of groups like the Bilderbergers, the Trilateral Commission, and MJ-12.

One species of weapon that is now being fielded by the military is of the "non-lethal" variety, aimed at incapacitating rather than killing the enemy.

Although research into "non-lethal" weapons has been conducted since at least 1976 in America, there have been enough leaks about the nature of this advanced technology that even the military is currently admitting to their intense interest in it. According to one commentator, Dan Goure of Washington's Center for Strategic and International Studies.

"The world is changing and our military's role is changing. The capabilities they have don't seem to match the new roles we see out there. There is a growing sense we need new tools."

In other words, more effective weapons, particularly for the cause of the moment, namely the suppression of international terrorism. This point should be carefully considered, since with the passing of the Anti-Terrorism Act, there is little difference between a bomb-throwing radical and a person who wishes to merely keep a weapon in the house for self-defense.

In 1994 Defense Under Secretary John Deutch authorized the latest step into the development of this new generation of "non-lethal" weapons. Deutch issued orders for a Pentagon team to go into high gear on the development of "NLW"s, tapping

UFOS, TESLA, AND AREA 51 by Commander X

Frank Kendall, the Pentagon director of tactical systems, to head up the team.

A "NON-LETHAL" CHECKLIST

Here are a few of the exotic weapons that the Pentagon has come up with since they have begun research in this area, and this is not even a complete listing of the types of weapons currently developed or being researched:

Disorientation lasers

Aerosol sprays that turn metal brittle

Sound generators that are so loud they produce unbearable pain

Strobe lights that cause nausea

Gases that are not deadly, but that immobilize the enemy

Flash guns that blind

Electromagnetic zap guns

Ultrasound beams that are so strong they can shatter buildings, as well as the internal organs of the enemy

Infra-red transmitters that cause buildings to burst into flames

Supercaustics are acids that are millions of times more powerful than conventional acids

Sleep agents that can put entire armies to sleep

Extremely Low Frequency generators that can inject voices into a person's head or destroy the immune system

Laser beams that will explode eyeballs

UFOS, TESLA, AND AREA 51 by Commander X

A wide variety of hallucinogenic drugs for injection into water supplies

Isotropic radiators are conventional weapons that fire omnidirectional laser rounds that dazzle people or optical equipment

Non-nuclear electromagnetic pulses produce huge amounts of power that can explode ammunition dumps or paralyze electronic systems

And many, many more advanced technologies...

NON-LETHALS IN ACTION

Some recent instances in which NLWs were used include Operation Desert Storm, when the U.S. Navy fired cruise missiles that dumped millions of infinitesimal carbon filaments on the populace, in order to disable machinery (or so we are told was the purpose). NLWs have also been employed in Panama and Grenada, according to Pentagon sources.

Like portable, small versions of the HAARP array, one of the most destructive of the new technologies is undoubtably electromagnetic beam weapons, which were recently tested in the Gulf of Mexico, to the objection of environmentalists. These weapons could bring entire nations to their knees, and we can only hope that it is not our nation.

Again, while it may be thought that these weapons will only be used on international terrorists, that is obviously not the case. For instance, Russian mind control systems were debated being used during the standoff at the Branch Davidian compound, according to reports in the Village Voice.

And FBI Agent Mike Toulouse testified at the Waco trial about FBI speakers that were set up around the compound, broadcasting high energy noise (including the horrendous sound of rabbits being killed) in order to drive up the stress levels of the people inside the compound.

Is there any doubt that this non-lethal arsenal will be used against people that the State considers 'internal enemies'?

UFOS, TESLA, AND AREA 51 by Commander X

If any confirmation is needed that high tech non-lethal weaponry exists in the arsenals of the United States military, the following article will surely provide it. Printed in the Spotlight newspaper, it is titled "Death Ray Fielded By U.S.," and authored by James Harber and Terry Dobbs. This information came to light during the U.S./Iraqi War.

For some Arab infantrymen sent into battle, their first look at advancing American troops may prove the last one. Behind a smoke screen of denials, the U.S. Army has begun deploying against Iraq the first working model of a dreaded secret weapon: a portable laser beam that will destroy the eyesight of enemy ground troops.

In recent months, the Pentagon has repeatedly denied that it was experimenting with a battlefield laser system that blinds people. But in a series of interview with confidential military sources, it has been found that behind the scenes a crash program to perfect such a weapon was assigned in 1986 to the Defense Advanced Research Projects Agency (DARPA), the Pentagon's leading science unit.

The laser gun sweeps advancing enemy troops with a beam intense enough to puncture and destroy their eyeballs even when they are not looking directly at the light source. It is now 'in the stage of test deployment,' a team of reporters has confirmed.

Code-named 'Project AOC' (Army Optical Countermeasures), the closely guarded design and development program has come up with a portable laser unit that 'would easily fit into your ordinary station wagon,' one senior Army scientist told this newspaper.

'This research is classified several grades above 'top secret,' warned this knowledgeable source, who urged that 'my identity must not even be hinted at' before he agreed to discuss this sensitive topic... 'This weapon -- what we used to call the 'death-ray laser' -- is now going to war, whether international treaties and humane considerations permit it or not,' revealed the veteran DARPA researcher. 'Since 1987, this has been our most urgently funded priority project. The result was a series of breakthroughs that led to a working prototype last year.'

UFOS, TESLA, AND AREA 51 by Commander X

According to this astute observer and other well-placed sources, the key problem confronting researchers was the beam power required and its unwieldy energy source.

'A high-frequency laser gun pulsating 10 joules of beam power 100 times per second will incinerate a human eye,' explained a consultant involved in the program's early stages. 'But to reach this intensity, it required cumbersome chemical power sources, the size of a 10-ton truck.'

(A joule is a term used to quantify energy. One joule equals 10 million ergs. Ten joules equals the work required to accelerate a quarter-pound weight to 100 feet per second.)

Concentrating some of its best staff scientists on the program, DARPA has developed what one source called a 'near-miraculous' compact solid-state power generator for the battlefield laser in less than a year.

'This brand-new weapon can now be taken into combat on an ordinary Army jeep, and of course aboard any helicopter,' confirmed a former DARPA program coordinator who now works as a private science consultant to the Pentagon...

'They want this inhuman and terrifying weapon to hit the enemy with the force of a shattering surprise,' says one veteran investigator for the House Armed Services Committee who asked not to be quoted by name. 'There are psychological studies indicating that the sudden terror of losing one's eyesight will panic troops more quickly than the deadliest gunfire.'

There are, moreover, indications that battlefield use of a blinding laser beam is in violation of international treaties on the conduct of military operations, congressional sources say.

Other advanced testing of weapons technology took place during the Iraqi war, including the use of radioactive bullets. According to Nexus magazine, of January/February 1992:

The continuing aftermath of the Gulf War leaves some of us wondering -- 'who were the good guys, and who were the bad guys?'

UFOS, TESLA, AND AREA 51 by Commander X

A press release from the Iraqi Embassy, Canberra, November 27, 1991, reports that apart from the existing list of atrocities committed by the UK and USA which includes burying thousands of Iraqis alive and deliberately bombing civilian targets; and allegedly using 'mind control' weaponry; we now discover that allied forces were using anti-tank shells containing depleted Uranium (U-238).

In other words radioactive bullets, designed to pierce armor were fired in the tens of thousands.

According to the International Commission on Radiological Protection, the tank ammunition alone would contain more than 50,000 pounds of uranium, enough material to cause more than 500,000 deaths. This does not count the bullets fired from aircraft.

To make matters worse, soldiers, construction workers, and mine-clearing experts working in the Gulf since the war, have not been fully informed of the risks posed by this uranium.

According to a UK newspaper, The Independent on Sunday, dated November 10, 1991, the Atomic Energy Report on the health risks involved has not been circulated to the workers cleaning up in Kuwait.

For additional knowledge of the purposes that the New World Order's non-lethal arsenal will be put to, recently reporter Harold Stockburger met a soldier who had personally encountered the plans of the Secret Government for aiming these capabilities at the populace of the U.S., rather than that of outlaw nations. The soldier requested anonymity, due to the explosive nature of his revelations. Here is Stockburger's report:

This young man told of how he entered Basic Training 1 and 1/2 weeks after the Oklahoma bombing and tells what he encountered during this period of basic training. The classes taken at Ft. Jackson, South Carolina were given to these men in the first three weeks they were there. The classes, described to us as being very intensive, lasted eight hours a day.

UFOS, TESLA, AND AREA 51 by Commander X

One of the first statements he heard, and one of the most startling was, 'If you'll read the clause in your contract, you might even have to fight domestic violence because of the terrorism and all these militias.'

Soldiers in class were then asked to define domestic disturbance by giving examples. One young soldier gave 'riots' as an example. Then they were asked to give an example of a large domestic disturbance. There were many examples given which were not what the instructor was wanting. Then one soldier stood and said, the 'Civil War.' The instructor then emphatically announced, 'That's correct!' He went on to explain how the Civil War was wrong, and how if it happened again, our military would be the 'Union' soldiers.

This instructor went on to prove his point by stating that 'certain states still fly the Confederate flag at their capitals, which represents disorder.'

The new soldiers were told how the Army is becoming a world policing organization. [This reporter] was told that at least 30% of the slides used during the training were showing U.S. soldiers wearing the U.N. blue helmets.

One of the more disturbing elements of the training centered on the rights of U.S. civilians. In one of the more memorable sessions the soldiers were told that, 'A lot of crazy Good Ol' Boys still want to carry guns.' 'Good Ol' Boys with guns' slang was used about 1,000 times, it was estimated by this soldier.

This was only the beginning. The instructors went on to warn the young recruits of the dangers of 'people stockpiling weapons,' and referred to them as a 'threat to the government.'

Then they were told, 'If there were to be another Oklahoma bombing type incident, it will be your responsibility to stop it.'

It was at this time that they were asked, 'Could you shoot at American civilians if they were up in arms against the government?'

There was a lot of discussion about riot suppression and militias. For two weeks there were classes dealing with terrorism and the Oklahoma bombing.

UFOS, TESLA, AND AREA 51 by Commander X

In October of this year, this new soldier went to AIT (Advanced Individual Training) in Ft. Sam Houston, Texas. What he received there was just as shocking, if not more so. This soldier just happened to be there when the situation arose with Michael New [A soldier who had refused to fight under the auspices of the United Nations, rather than the United States].

Because of the new situation the entire group was addressed by the Company Commander. He told the soldiers, 'We are part of a one world military. We'll be able to stop aggression no matter where it happens in the world. We will police the world. A lot of people are isolationists. Isolationists are right up there with racists and the KKK. They have no compassion for the rest of the world. They don't believe in a world police force.'

The Commander went on to say, 'World wars were fought to establish the U.N. Do everything you are ordered. You'll never be able to get a job in the U.S. again if you do this (like Michael New). He (New) will be court martialed. If you refuse, you will be court martialed and will never be able to work again in this country.'

Again, during this segment of training, they were asked, 'Could you fire on civilians who are up in arms against the government, like in another civil war? People who are stockpiling weapons are a threat against the government. Would you fire upon them?

This soldier also told this reporter that there were always soldiers hanging around after the class asking, 'What do you think about all this?' He said, 'Many of the soldiers felt these men were posing as just one of the guys so they could find out which of us were having problems with this, so they could turn us in.'

In reference to the political correctness of the Army, they were told the 'Army Life Magazine is now considered a political magazine. If you read anything that promotes a different view, you may get called up for it.'

One soldier got an article 15 for an altercation stemming over politics. He had a conservative, Christian view of national politics and was reported on by another soldier for some literature he was reading.

UFOS, TESLA, AND AREA 51 by Commander X

During these three weeks they were trained in 'how to go door-to-door, rounding up guns from SUSPECTED citizens.'

Soldiers who objected or complained were called Nazis or Rednecks, and harassed into going along.'

UFOS, TESLA, AND AREA 51 by Commander X

Here is photographic proof that the technology exists that enables the New World Order to control our most private thoughts and sensations. Such "implants" are in wide use even today.

CHAPTER TWELVE
MIND ARSENAL

Additional information on the technological capabilities of the New World Order is found in an anonymous document titled The Thousand Year Reich -- The First Fifty Years, which links early experimentation by the Nazis, the transplantation of Nazi scientists to the United States, and the researches of the U.S. government in recent years. Revealed here is the real cutting edge of experimentation into technologies far in advance of what has been admitted by the government, and many ramifications of this research can be imagined by the reader. In this article it is stated,

In the 1950's, several experiments were conducted with a telepathic sender and a receiver located a thousand miles apart. The only aspect new to the experiments was the inclusion of highly sensitive monitoring devices that recorded the minute energy flares on an electronic gauge as it left the sender and arrived at the receiver. The experiment only confirmed previous findings but one unheralded genius came up with a brilliant idea. The experiment was repeated but, instead of having the sender concentrate on a human receiver, he attempted to send his telepathic message directly to the monitoring equipment. The scientists were amazed when, at the appointed time, a thousand miles from the sender, the needle on the monitoring equipment moved. The machine could not understand the message, only that a message had been received, so the experiment was repeated but this time, instead of sending a speech or thought pattern, the sender projected a series of energy pulses in Morse Code. The gauge registered the dots and dashes of the message and easily translated them into the written word.

It was a monumental discovery; the first mind/machine link. By 1965, this unpublished science had so advanced as to be applied as the communication system for an ultra-secret intelligence group in the United States. One of the world's largest computers (the same Hewlett-Packard model as used by the General Services Administration to run the Federal government) was set up and programmed to receive the telepathically communicated reports of thousands of intelligence agents in the field. It is the quickest, most efficient, untappable communication system in the world today; linking the life experiences of thousands of covert operatives directly into the computer data banks that, at a moment's notice, can regurgitate the sum total of their knowledge on anything or anyone so requested. The organization exists today with headquarters in California's Silicon Valley and branches throughout the world that operate in conjunction with acupuncture clinics as conditioning centers for the operatives. It was one of the best kept secrets.

The science of telepathic communication has not escaped the interest of the Central Intelligence Agency, which recently funded several studies in parapsychology in a joint venture with the National Security Agency in an effort to establish a super-secret nation defense system utilizing the 'energy influences' of thousands of operatives to jam the directional guidance systems of incoming enemy missiles [or discoid craft]. It is a plausible defense.

Likewise, the CIA reports indicate that as early as 1972, the characteristically austere Soviet Union had allocated an annual budget of over $20 million for their research in this field.

PSYCHOTRONICS

The following anonymous report issued to researchers on the Internet offers additional clues as to the purpose of non-lethal weaponry:

All major news media have recently proclaimed that false memory syndrome is likely to be the next big problem in the mental health professions for the 90s. Do not buy into these particular news releases, which have been planted by clandestine spin doctors for the express purpose of covering-up a massive ongoing abuse of U.S. citizens at the hands of a variety of Federal level agencies who are deploying a variety of so-called "Novelty effects" which are classified as "Non-lethal" yet may still cause injury, health problems, and post traumatic stress symptoms. This non-lethal technology has been deployed for nearly forty years, and is used for many purposes -- some of which are suggested below.

MASER, ma' zer (microwave amplification by stimulated emission of radiation), a device that produces highly stable electromagnetic waves by harnessing the natural oscillations of an atomic or molecular system. The Maser devices have been a staple of "our" National Security States of America since before 1955, the

UFOS, TESLA, AND AREA 51 by Commander X

year that the maser was hailed as an important scientific achievement in the realm of psychological warfare

These devices have been deployed through the following Federal agencies: FBI, NSA, NRO, CIA, Army Intelligence, Navy Intelligence, Department of Energy, Defense Investigative Services, DIA, Justice Department, Department of Treasury, BATF, and probably others which I am not aware of the in the Federal Reserve System and State Department. Deployment of this technology is a distinctly federal phenomenon!

The so-called "Secret Government" will not allow mainstream media to speak or write the truth about the matter, and will deny this is the case at every turn. Masers are LPM (low power microwave) directed energy weapons which are capable of causing the projection of auditory effects (either hearing "voices" or subliminal suggestion) into the occipital cortex. There is no way to shield oneself from this form of extremely high frequency directed energy -- no shielding is adequate! One's only hope of protection is to be aware of the abuse as external to self, i.e. not coming from one's own imagination or delusions, but invasive projections designed to effect mind control.

False Memory Syndrome is a scapegoat created by a consortium of Federal "spin doctors" bent on negating the believability and viability of the more than 12,000 unwitting citizens who have been on the receiving end of this technology. Since most persons on the receiving end have also been the types the government considers "enemies of the State" to begin with, the perpetrators have been allowed to use their victims sexually as well. That's right, rape, but a form of rape that is not available to the average rapist who hasn't access to the so-called "psychotronic" novelty effects weapons. Psychotronics allows the perpetrators to sexually abuse and rape their victims and yet still have the victims enjoy and experience full sexual response. Yup, the victims have had their will power severed and give their abusers more pleasure than any unwilling non-consensual partner ever could!

The mental health professions are currently swallowing this bogus media barrage on False Memory hook, line and sinker -- and are preparing to ban all repressed memory therapies (hypnotherapies) as the main cause of these so-called "Lies of the Mind". The 12,000 victims of psychotronic abuse will then be dealt with as mere schizophrenics, just like they always have been for over 40 years! The media planted stories about False Memory syndrome will help "explain away" the memories of Alien Abduction/Ritual Satanic Abuse which are so prevalent in case of government psychotronic abuse. The "Satanic" abuse being ritual abuse by the Secret Government who control the Federalist World Government of the New World Order/United Nations juggernaut! Are their eyes and ears really benevolent? Not! One of the main reasons that this form of atrocity is being so deftly covered up is due to the consistent abuse, sexual and otherwise, of the persons involved in the government's long standing "genetics vs. heredity" eugenics and genetics experiments which the imported Project Paperclip Nazi scientists have helped our government with since WWII!

These same covert arms of government are the ones who have coined the term "screen memories" to describe the obfuscational memories impressed by the abusers themselves! They must at all costs disguise their abuse in order to continue experimentation with psychotronics, and have given a fancy neutral name to their enforced amnesia which is designed to disguise abuse during domestic operations. Allowing the perpetrators their sexual reward is part and parcel of payment to the covert operatives in the field. Again, screen memories are enforced amnesia. Since these Nazi inspired scientists must perform medical tests during the abuse event, such as implantation of bio-telemetric tracking devices into nasal cavities and ear canals, retrieval of gamete samples/tissue samples, they use alien abduction as a screen memory, inspiring terror in the populace they abuse, while the clueless "sheeple" will not believe in extra-terrestrials from space, those on the receiving end of the abuse are often convinced of space alien abuse or satanic abuse. This con job is fooling even the best of mental health professionals! The atrocities are so outrageously unbelievable that justice will never ever be brought to bear on this most secret activity of "our" Secret Government. The Biotelemetric implants are used to locate the victims via the Global Positioning System (GPS) satellites, they cannot escape or hide from their abusers.

Victims are typically selected for a variety of reasons: lifelong eugenics/genetic guinea pigs, children of secular humanist society members, enemies of the state, suspected spies, gypsy types (includes hippies, beats, deadheads), war protestors, civil rights activists, suspected terrorists, persons who use any drugs which the state cannot tax, feminists, homosexuals, socialists who differ from New World Order objectives, firearms owners, or anyone doing anything which irritates the powers that be.

UFOS, TESLA, AND AREA 51 by Commander X

The Federalists have a special name for the select few abuse victims that they wish to control for clandestine objects -- unit sleepers -- who can be made into "Manchurian Candidates" like the crazed gunman with killing spree madness that are used in Post Office shootings, schoolyard shootings, Luby's Cafeteria shootings, New Jersey train shootings. They would have you believe in the old "lone crazed gunman" scenario in order to convince the populace that they must be disarmed for the national good. They will succeed.

Readers will note that this past Christmas season was one of the most violent seasons in history with regards to firearms mayhem. It is being engineered. The end justifies the means when it comes to disarming the populace as far as They are concerned. It matters not that they are really responsible for the atrocities attributed to crazed lone gunmen, since they believe the benefits will only come after they disarm the law abiding sheeple, leaving criminals armed in order to foster anarchy which must eventually result in bringing in the U.N. to perform peace-keeping duties within "our" nations borders. Martial law will be declared, and the Constitutional rights suspended under this nation "drug and crime" emergency.

The Fox Network is currently using the X-Files to reveal the government's position on these issues. However, the extraterrestrial hypothesis is the only official accepted position. You will not see Special Agent Fox Mulder uncovering evidence of government abuse of citizens, except when the government is deploying crashed saucer retrieval teams!

Get a clue, sheeple, your time is short!

UFOS, TESLA, AREA 5l by Commander X

Disk-shaped craft constructed on other planets are now being kept in workng order by heavily armed military personnel at several locations usually located underground. Several abductees have been examined in alien instillations far below the surface.

CHAPTER THIRTEEN
MIND CONTROL

Experimentation into mind control has been going on literally since the beginnings of recorded history, although its existence has been very carefully excluded from the public mentality -- as so much has. If people know that control is going on, it is so much more difficult for the Secret Government to accomplish.

CIA MEMO

Here is a previously Top Secret CIA memorandum that tells part of the story and substantiates the existence of mind control projects conducted by the American government:

22 November 1961

MEMORANDUM FOR: THE RECORD
SUBJECT: Project MKULTRA, Subproject No. 94

1. The purpose of this subproject is to provide a continuation of activities in selected species of animals. Miniaturized stimulating electrode implants in specific brain center areas will be utilized.

2. Initial biological work on techniques and brain locations essential to providing conditioning and control of animals has been completed. The feasibility of remote control of activities in several species of animals has been demonstrated. The present investigations are directed toward improvement of techniques and will provide a precise mapping of the useful brain centers in selected species. The ultimate objective of this research is to prove an understanding of the mechanisms involved in the directional control of animals and to provide practical systems suitable for [Censored in the original] application.

UFOS, TESLA, AND AREA 51 by Commander X

3. [Censored in the original] will function as cutout and cover for this subproject. The research and development will be conducted in facilities provided by [Censored] who will submit to [Censored] a summary of accounting of funds at the conclusion of this program. Any unused funds remaining at that time will be returned to [Censored].

4. The cost of the program for a period of one year beginning 1 December 1961 is estimated to be $55,222.90. To this sum must be added $2,208.92 which represents the 4% service charge due to [Censored]. The total cost of the project for a period of one year will, therefore, not exceed $57,431.82. Charges should be made against allotment number 2125-1390-3902.

5. It is not anticipated that any permanent equipment will be needed for this program. Documentation and accounting for travel expenses which are reimbursable by [Censored] will conform to the accepted practices of that organization.

6. The requirement for a semi-annual informal accounting on the part of the principal is waived.

7. All personnel connected with the planning and monitoring of this program possess TOP SECRET approval. the project will be unclassified after it leaves [Censored].
3   Signed [Censored], Chief TSD/Research Branch

Approved for Obligation of funds by [Censored]

Date [unintelligible]

Attachment: Proposal and Budget

Distribution: Orig. only.

The "selected species of animals" which the above secret document refers to (with the stated intention of injection with a mind control device) almost certainly must include humans. The reason that this is so? If humans were not involved as test subjects, then it is highly unlikely that this project would have had a Top Secret designation. Also, there is no shortage of accounts of testing upon humans using these exact technologies in the annals of MKULTRA, some of which will be detailed in the pages that follow.

MIND CONTROL IN HISTORY

Samuel Chavkin, the author of the Mind Stealers, summed up the issue of government mind control very well in the following letter to the editor, published in the New York Times:

Subjecting people to experiments, in most instances without knowledge of the risks involved and without their consent has been a continuing practice by government agencies. A January 5 news article discusses the difficulty the Central Intelligence Agency is having finding records of the experiments.

However, at a Senate hearing on Aug. 3, 1977, Admiral Stansfield Turner, former CIA director, disclosed that the agency had been conducting brainwashing experiments on countless Americans -- prisoners, mentally ill patients, cancer patients, and even unwitting patrons at bars in New York, San Francisco and other cities. Some were drugged with LSD and other psychotropic agents.

This was the cold war period, when the focus was on spying and counter-spying. Thus the main objective of this mammoth CIA effort, which cost the taxpayers at least $25 million, was to program the experimental subject to do the programmer's bidding, even if it would lead to the subject's destruction.

As you reported August 2, 1977, a CIA memorandum of Jan. 25, 1952 asked, 'whether it was possible to get control of an individual to the point where he will do our bidding against his will and even against such fundamental laws of nature as self-preservation.'

Mind control and behavior modification experiments in this period also became the underpinnings for a 'medical' approach to stem the rise of social disquiet following the murder of the Rev. Dr. Martin Luther King, Jr. Hundreds of thousands of Dr. King's followers were out in the streets throughout the United States demanding that their civil rights be recognized and that Dr. King's assassins be brought to justice. Many protests led to violent confrontations with the police.

UFOS, TESLA, AND AREA 51 by Commander X

Two physicians from Harvard, Dr. Frank E. Ervin, a neuropsychiatrist, and Dr. Vernon H. Mark, a neurosurgeon, in a letter to the journal of the American Medical Association, proposed a surgical strategy to resolve such conflicts. In their view, protestors who resisted police control were suffering from 'brain dysfunction,' a condition, they said, that could be remedied by psychosurgery.

They proposed implantation of very thin electrodes in the amygdala region of the brain where 'bad' brain cells, presumed to be associated with violent behavior, would be burned out with an electrical charge.

Despite an angry outcry from many physicians who charged that this was in effect a return to the discredited lobotomy operations used on shell-shocked soldiers following World War II, law enforcement authorities welcomed the approach. Especially impressed with psychosurgery was Ronald Reagan, then Governor of California, who was ready to allocate $1 million to set up a 'violence reduction center.'

Without much further ado, psychosurgery got underway in the Vacaville penitentiary in California; at Atmoree State Prison in Birmingham, Ala., where 50 such operations were performed, and in other prisons. The Veterans Administration used psychosurgery in its hospitals in Durham, N.C.; Long Beach, Calif., Minneapolis and Syracuse. As a result, many prisoner guinea pigs entered into a semi-vegetable state of mind.

Psychosurgeries were finally halted when civil libertarians and the Congressional Black Caucus denounced them as racist, since most of the prison population was made up of Afro-Americans and other minorities.

OTHER METHODS OF MIND CONTROL

There have been many, many methods of mind control practiced in the world, with the focus of these operations taking place in the United States, always the cutting edge when it has come to behavioral experimentation.

Beginning our overview of this technology with what took place in 1947, experimentation into mind control by the use of drugs was run at Bethesda Naval

UFOS, TESLA, AND AREA 51 by Commander X

Hospital under the direction of Dr. Gaefsky. This project continued until 1972, although there may have been offshoots that were launched under different names.

CIA reports admit that the purpose of the drug experimentation was to isolate drugs that would facilitate psychological entry and the control of people, and this goal certainly has not changed in these researches sponsored by the U.S., but ultimately by the Secret Government.

The mind controllers would soon see, however, that drugs were a rather primitive form of control when it came to human beings, and that far more fine-cut techniques were envisionable. As new forms of electronics and electromagnetic broadcasting were developed, American intelligence agencies were in the forefront in their usage of these technologies in controlling peoples' minds. The following is a report on varied forms of mind control, written by leading researcher Val Valerian:

The RHIC-EDOM file was a 350 page document prepared by the CIA immediately after the murder of JFK. The report described a way of turning men into electronically controlled robots that were programmed to kill on demand.

In the RHIC phase, the individual was put into a trance and given suggestions that were activated in one or more levels by key words or tones.

In the EDOM phase, which was an acronym for Electronic Dissolution of Memory, the memory of the individual was affected to either eliminate or alter the memory of events that the individual was involved in. By electronically jamming the brain, the existing acetylcholine creates static which blocks both
sights and sound. This method can be used to either block/erase the memory, or to slow it down so that events seem to happen after they actually have occurred...

The use of Very Low Frequency Sound or Ultrasonics has been well documented. VLF sound and Ultrasonics can affect both the electrical behavior within the brain and the actual brain tissue.

In the current phase of international VLF warfare, pulses of 7-12 Hz are bounced off the 8 Hz ionospheric envelope around the Earth. Within these pulses are entrained bizarre and aberrant patterns that produce equivalent behavior in humans.

UFOS, TESLA, AND AREA 51 by Commander X

In both California and Washington state there are extensive problems with the results of this type of activity.

In 1961, the University of Illinois did experiments on ultrasonic research that eventually got into the hands of the military industrial complex...

The experiments done in 1961 by Drs. W. Fry and R. Meyers used focused ultrasonics to make brain lesions of a controlled size. Their research demonstrated the advantage of ultrasonics over psychosurgical techniques which implanted electrodes in the brain.

In 1963, Dr. Peter Lindstrom at the University of Pittsburgh used a single unfocused sonic beam to destroy fiber tracts in the brain without damaging the nerve cells next to them. This pre-frontal sonic treatment replaced lobotomy...

EXTREMELY LOW FREQUENCY

The use of extremely low frequency (ELF) generation as a weapon is documented as far back as the 1960 (at least in declassified documents, although it is quite possible, according to some accounts, that this research began much earlier, after the American government gained access to the researches of Nikola Tesla.

The American military was very careful to keep themselves distanced from the development of these "beam weapons," and "mind control projectors," but they were willing to talk about Russian experiments in attacking human beings, thus justifying their own researches to funding committees. The Russian mind control capability was viewed as a major Cold War threat, thus the American military research community was mobilized to counteract this threat. What was not mentioned was that we had always been ahead of the Russians in this race, not for the Moon, but for the mastery of men's minds.

Not to say that the Russians weren't involved in this sort of research, too. One such program was when the Russians irradiated the American Embassy in Moscow through the broadcasting of electromagnetic waves, this assault beginning about 1967. The Embassy employees suffered a variety of horrendous effects from their exposure to the electromagnetic broadcasts, including extreme mood swings, higher incidence of sickness and disease, and even bleeding from the eyes. The head of the

UFOS, TESLA, AND AREA 51 by Commander X

American Embassy in Russia at the time developed a rare blood disease which was probably attributable to the electromagnetic assault.

CONTROL TODAY

For a glimpse into the true capabilities of mind control and related technologies as they are currently being practiced by the military in secret bases around the world, we would do well to consult the classic research essay "The Dulce Base," by Jason Bishop III, which blew the cover off of alien/human collaboration several years ago. In this excerpt, Bishop is referring to experiments allegedly taking place in the huge underground base located at Dulce, New Mexico:

The Dulce Base has studied mind control implants; Bio-PSi units; ELF devices capable of mood, sleep, and heartbeat control; etc.

DARPA (Defense Advanced Research Projects Agency) is using these technologies to manipulate people. They establish the projects, set priorities, coordinate efforts and guide the many participants in these undertakings. Related projects are studied at Sandia Base by the Jason Group of 55 scientists. They have secretly harnessed the dark side of technology and hidden the beneficial technology from the public.

The studies on level four at Dulce include Human-Aura research, as well as all aspects of dream, hypnosis, telepathy, etc. They know how to manipulate the bioplasmic body of man. They can lower your heart beat, with deep sleep delta waves, induce a static shock, then re-program, via a brain-computer link. They can introduce data and programmed reactions into your mind (information impregnation -- the "dream library").

We are entering an era of the technologicalization of psychic powers. The development of techniques to enhance man/machines; Psi-war; EDOM (Electronic Dissolution of Memory); RHIC (Radio-Hypnotic Intra-Cerebral Control); and various forms of behavior control via chemical agents, ultrasonics, optical and other EM radiations. The physics of consciousness...

We consult leading researcher Val Valerian further for additional information on the capabilities of mind control as employed by the Secret Government:

UFOS, TESLA, AND AREA 51 by Commander X

A research and development team at the Space and Biology laboratory of the University of California at the Los Angeles Brain Research Institute found a way to stimulate the brain by creating an electrical field completely outside the head. Dr. W. Ross Adey stimulated the brain with electric pulse levels which were far below those thought to be effectual in the old implanting technique.

Ten years before, Dr. Delgado described a society kept under control by electronic brain manipulation. By the late 1960s, remote control of the brain was well on its way to being realized.

In 1975, a primitive "mind-reading machine" was tested at the Stanford Research Institute. The machine monitored silent thoughts. The machine was the creation of psychologist Lawrence Pinneo and computer experts Daniel Wold and David Hall. Their originally stated goal was to put a computer programmer into direct contact with a computer. The concept promised benefits for the physically handicapped, but as with everything new, the military applications quickly came into view.

In 1976, a young scientist at the Rockefeller University named Dr. Adam Reed made another conceptual proposal. He said that "by 1996 it will be feasible to encode and transmit brain waves from a small device implanted inside the skull."

This development promised transmission of data from computers directly into the brain, which parallels the processes that some of the alien groups use in their inculcation (forced learning) process that they impose on androids or humanoids that they use as workers. This process was eventually worked out and made the way clear for the development of **RHIC-EDOM** (Radio Hypnotic Intra-Cerebral Control - Electronic Dissolution of Memory) and allied applications.

Intelligence forces have already developed remote controlled men who have no memory of their programming, but will perform a preassigned task when exposed to an audio or visual cue. These men, who are essentially cyborgs (altered and controlled humans) are far less expensive than robots. They are also expendable. There is now sufficient evidence to indicate that the assassinations of John and Robert Kennedy, as well as Martin L. King, were carried out by individuals who were programmed with **RHIC-EDOM** techniques by intelligence forces.

UFOS, TESLA, AND AREA 51 by Commander X

The court system, in all but a few cases, has retained the position that criminal behavior cannot be induced under hypnotic methods. If the intelligence forces in the United States and elsewhere stand outside the law that applies to the rest of us, then it doesn't take a lot of imagination to see where events will lead if we are not vigilant.

STATEMENT OF A VICTIM

To really understand the horror of mind control, one must talk to one who has experienced it. Although many victims of mind control have come forward to protest what has happened to them, some reports are outstanding in terms of their objectivity and understanding of what is going on. The following is a portion of a report by Glen E. Nichols:

I am a victim of mind control. I have been mind controlled for my entire life of 46 years. I will try to briefly summarize what I remember and why I am writing.

I have been mind controlled with the typical methods of behavior modification, hypnosis, drugs, electric shock, torture, and lies. I believe that most people who understand the methods of the secret societies would call me a Greek slave. I can recall the perpetrators using the above methods on me when I was younger than three years old. As you know, the methods cause amnesia for the conditioning and psychological dissociation of the reality that the person is a slave.

Over the years the above methods were frequently used on me, at times daily. Sometime I would realize my circumstance because some good people would awaken me and explain what was being done to me. When I would begin to realize what was being done to me, the perpetrators would find me and intensify their efforts. They often used torture, ether, memory blocking drugs, and electric shock to my head to erase my memories. This would cause me to forget what the good citizens told me.

Apparently, the above very strong methods of mind control were not enough for the perpetrators. I now remember that when I was about 12 years old in 1959, these evil people inserted a temporary radio receiver in my ear. I was told that I had a hearing problem and I had to wear a hearing aid.

UFOS, TESLA, AND AREA 51 by Commander X

Apparently, some people around me objected, and they stopped that tactic for a short time. When I was about 15 years old in 1962, they had miniaturized radio receivers sufficiently that they could insert one in my ear canal without me realizing what they had done. Then they would use very low volume subliminal messages to control me. They would give me post-hypnotic suggestions that I had ear infections and I must not clean inside my ear canals or I would suffer permanent hearing loss.

Occasionally, I would clean inside my ear and find a metal device, that would be explained away by those around me. The realizations of devices in my ears were then always erased with torture, drugs, and electricity.

The mind control efforts were very intense. I was oblivious to many events around me, and certainly to the reality I was being completely controlled. I had thousands of days and experiences simply erased by their techniques. Occasionally, with some people's help I would briefly break free from their control. This would enrage them.

When I was about 20 years old in 1967, they surgically implanted a miniature radio receiver and transmitter behind my right ear canal and next to my ear drum. I was told I had hurt my ear playing football.

Over the next 20 years the receiver-transmitter was replaced a couple of times and another one was implanted behind the other ear. I have several very fine scars above and behind both ears.

Apparently the perpetrators would not only broadcast subliminal commands to me, but had the ability to listen to my conversations, and to electronically track and locate me.

For some reason, in addition to the surgically implanted radio receiver-transmitters, at times additional miniature receivers were also inserted into my ear canals. I can recall having one found in my ear by a Cal State Northridge health center doctor in 1971. I also found another one in my ear about 1988. Of course they would immediately erase my memory of what I had found. I assume there have been other times that I have not yet recalled.

UFOS, TESLA, AND AREA 51 by Commander X

Why would there be miniature receivers in my ear canals in addition to the surgically implanted ones behind my ear canals? I can only guess. Perhaps different organizations were trying to control me.

Maybe people were monitoring the frequency of the implants, so the perpetrators would send extra sensitive commands with the new devices. I really don't know. I do know that any realization of what they had done to me was quickly erased from my memory. I lived for about 42 years not realizing that I am a victim and they have turned me into a human robot.

There have been many other devices. I can recall having an electrode inserted into my frontal brain through my nostril when I was about 33 years old in 1980. I experienced intense pain, confusion, and disequilibrium. I was told I had a sinus infection. I explored the source of my pain and found a small bulb stuck to the roof of my nasal cavity. The doctors I contacted said it was an infection and not to touch it. I persisted and removed it myself with tweezers.

I looked like a two-pronged electrode, with two sharp wires stuck up into my brain with the bulb hanging down. I took it to the FBI. They told me it was a transistor that had probably fallen out of an automobile. Then my memory of the event was erased for several years until recently.

INTERVIEW WITH AN INSIDER

The following is an interview with a CIA employee, who for obvious reasons, must remain anonymous. This interview was published in Leading Edge magazine, and we are pleased to be able to present this insider's view to you of what is going on in the Secret Government:

Q: A career officer doesn't need to be debriefed by mind control, does he?

A: You want to bet? This debriefing is done in such a way, in many causes, to cause actual memory damage.

Q: What about how the government works?

UFOS, TESLA, AND AREA 51 by Commander X

A: Don't kid yourself. This country is controlled by the Pentagon. All major decisions are made by the military. The CIA's just the whipping boy. NSA are the ones who have the hit teams. Look into their records -- you won't find a thing. Look into their budget -- you can't. The CIA is just a figurehead, but as far as intelligence goes, the NSA's far superior to them -- far in advance in the 'black arts.' The CIA gets blamed for what the NSA does. NSA is far more vicious and far more accomplished in its operations. The American people are kept in ignorance about this -- they should be, too.

Q: What you're saying is that the military is more dangerous to our democracy than the CIA or other intelligence groups?

A: The CIA gathers the information, but the military heads the show.

Q: What you are suggesting, I guess, is that there is an invisible coup d'etat which has occurred in the United States?

A: OK. There is a group of about eighteen or twenty people running this country. They have not been elected. The elected people are only figureheads for these guys who have a lot more power than even the President of the United States.

Q: You mean the President is powerless?

A: Not exactly powerless. He has the power to make decisions on what is presented to him. The intelligence agencies tell him only what they want to tell him. Think of the Pentagon papers. It's public knowledge that the CIA has falsified documents and done a host of other things. You have to wonder at American stupidity. What people don't know is that the global corporations have their own version of the CIA. Where they don't interface with the CIA, they have their own organizations -- all CIA-trained. They also have double agents inside the CIA who are loyal to those corporations.

Q: What do you know about the use of pain-drug hypnosis?

UFOS, TESLA, AND AREA 51 by Commander X

A: They use several different things. I've seen guys coming back with blanks only in certain places of their memory. They use hypnosis and hypnotic drugs. They also use electronic manipulation of the brain. When they use hypnosis, they'll be using at the same time a set of earphones which repeat 'You do not know this or that' over and over. They turn on the sonics at the same time, and the electrical patterns which give you memory are scrambled. You can't hear the ultrasonics and you can't feel it, unless they leave it on and then it boils your gray matter.

Q: I thought that research had stopped on ultrasonics?

A: Yeah, the research has stopped. They've gone operational. It ain't research any more. They know how to do it. Our Constitution doesn't permit us to do much that is legal.

Q: Do the police use mind control?

A: At the highest levels, yes. The FBI uses it, and they give a lot of help to local police. Let me tell you something: the cheapest commodity in the world is human beings.

Q: What about conspiracies?

A: All you hear about are left-wing conspiracies to overthrow our government. You never hear about right-wing conspiracies. Well, some of these right-wing groups are far more dangerous than the left-wing groups. The right wing is usually retired military. If the right-wing took over right now there would be a military dictatorship. We've got one right now, but it ain't overt, it's subtle.

Q: You mean those 20 men you were talking about?

A: Yeah... if the people of this country actually knew that, they would say 'no' the next time they were asked to go to Vietnam. We need the people behind us to fight a war, and if they knew the true facts, who's running things, there wouldn't be the following we'd need to defend the country. that fact alone keeps the sham of politics and 'free elections' going.

UFOS, TESLA, AND AREA 51 by Commander X

The American people, like most people, have to feel that they have some right in what they do, that they're the good guys. This is the reason we have never lost a war and have never won a peace. You've got to maintain the sham of freedom, no matter what. It wouldn't make any difference what party is in charge or whether it was the elected government or what is called the cryptocracy [or the Secret Government] that was running it; from an operational sense, the government would operate as it presently is.

UFOS, TESLA, AND AREA 51 by Commander X

Unmarked black helicopters, heavily armed military personnel and a supposed "invasion" by aliens could well turn out to be a plot by the New World Order to negate the U.S. Constitution.

CHAPTER FOURTEEN
ALIEN MIND CONTROL

In case you think that all mind control is sponsored by the CIA or another earthly intelligence agency, think again. It should be noted that implants are often injected into humans during extraterrestrial abduction. This has been verified in account after account by abducted individuals, and I have recounted my own experience with this in the extensive Commander X Files (Available for $40.00 from Global Communications, P.O. Box 753, New Brunswick, NJ 08903).

MIND CONTROL BROADCASTS

Here is a recent update on another form of alien mind control technology from Lorne M. Goldfader, Director of the U.F.O. Research Institute of Canada:

The "Greys" have an interesting yet frightening device in their arsenal of behavior modification weapons, which they use against certain Earth humans. They are geniuses compared to our own government scientists in the use of mind control. There exists an instrument which is organically grown to specifications, genetically engineered, brain wave sensitive, spheroid in shape, very lightweight and small enough to cup inside a child's hand, then positioned within 2-15 inches of a "controller's" forehead, thoughts can be electronically focused, magnified and projected into the mind of any targeted Earth human.

The usual and routinely anticipated outcome is to render the quarry ill along with induced headaches or severe stomach cramps and in some cases increased heart rates leading to artificially induced attacks and/or death. The illness is often concurrent timewise with an abductee's positive contact with the gentle and kind Federated species and is designed to implant into the subconscious mind the idea that the illness is alien induced. Since most human Earth contacts are not able to discern the originating source, they assume that all of their extraterrestrial contact is bad, and cease all UFO research activities. Sometimes individuals, if they are viewed as a future potential threat are brought directly into a craft lab, sometimes with their siblings, and vibrated violently on a table or against a wall, along with the use of sound waves and in conjunction with implanted ideas and memory block techniques are literally forced to avoid certain areas of growth, educational and vocational pursuits in order to circumvent certain future potential events from occurring which could, if not checked, lead to further evolution of mankind and therefore become a danger to Grey objectives.

Fortunately, some individuals have been given biological and microscopic detection and reflection systems in their bodies by the "others." The implant absorbs psionic energy sources. In conjunction with the conscious will and intent of the abductee, it reflects the waveforms back to the thought control warrior. This is a risk that the Greys are willing to take since the chances of originating source detection are almost nil at this stage of Earth human evolution and awareness. However, it is also their Achilles Heel

There is a law which allows direct intervention by the good Federated species when humans are under attack in such a manner without their conscious knowledge. An attached condition of this law is that free will choice must be exercised by the victim involved, and in a cooperative fashion (most often subconsciously). Increased awareness facilitates intercession.

If you suspect an attack is in progress (it can last anywhere from 2-5 days) and you display all of the symptoms but are unable to defend yourself or find the resources to do so, try to mingle with a large crowd of people (to wash out your brainwave trace signals). Get away from the area where you are having a problem. Get a magnetic map and go to a LOW intensity area of field strength (the Greys sometimes manipulate surrounding available energy resources as offensive weaponry).

However, the best and by far the most effective method is to mentally construct a shield of protective light around your body. The electromagnetic frequencies and nerve impulses of your brain will respond to these thoughts in a material sense. Imagine that the armament possesses a reflective surface so that anything coming your way bounces off. Now, direct this incoming energy to travel exactly in reverse back to its originating source and order it (with 100 per cent conviction) to fulfill its intended purpose on the transmitter. Ask for help from the highest levels of intelligence and love which is your greatest defense. There are always forces on your side. If you keep working for the good you may become like them some day.

Although you may be tagged and marked and tracked, remember that **YOU ARE NEVER ALONE** and you do not have to be a victim. All Greys respect mental strength. Not all extraterrestrial Grey species are bad. There are many variations of lineage. Our reaction and interpretation of their psychology and actions determines their strategies. Adaptation and integration into their mode of thought and culture is of prime importance to them. Do not conform and you will be punished. Our societies here on Earth do exactly the same with our own people so it is a matter of perspective.

This particular Grey species does not understand free will choice because they exist as a hive unit mentality, more genetically insect than human. To them, individuality and unpredictable emotion is dangerous --better to be bred out. They have much to learn from us -- especially the "elite" of the more cerebral Grey alien scientists. Only through mental and spiritual strength as well as cooperative abductee endeavors can we repel the invaders or at least gain their respect enough to be able to establish a two way communication process that may bridge the gap between us.

UFOS, TESLA, AND AREA 51 by Commander X

In parts of Puerto Rico ETs apparently walk about without even trying to hide the main thrust of their activities. They have been observed carrying out a variety of tasks.

UFOS, TESLA, AND AREA 51 by Commander X

CHAPTER FIFTEEN
FAR REACHES OF CONTROL

Exposing deeper levels of New World Order manipulation and control, many have heard of the infamous Philadelphia Experiment, along with the less-well-known Phoenix Project and the horrendous experimentation which took place at Montauk, New York beginning in the 1950s. These have been well-documented in other books, although it may be said that only the tip of the iceberg of this experimentation has been exposed at this date.

DARK BACKGROUND

An anonymous government scientist provides further material which challenges credibility due to its astounding implications in the area of government and alien mind control and other experimentation. The following interview has been circulated clandestinely among researchers and opens up entirely new levels of understanding about secret technology and the plans of the Secret Government.

Certain technical materials have been excerpted from this interview, for the sake of clarity for the non-technical reader.

Q: Let's start with the Phoenix Project.

A: It was a project that evolved out of the Philadelphia Project. It was a project that the Navy did in the 1930's and 1940's in an attempt to make the ships invisible. They threw the switch one eventful day and the ship went into hyperspace. They had all sorts of problems with the people on the boat. It was a huge success as well as a huge failure -- then they shelved it. Around 1947 it was decided to reactivate the project and it was moved to Brookhaven National Laboratories with Dr. N. and his associates.

UFOS, TESLA, AND AREA 51 by Commander X

Out of Phoenix 1 came Stealth technology, which I cannot talk about because of my job. It also produced all sorts of energetic little toys like the radiosonde.

Q: What is a radiosonde?

A: Well, in all appearances it was a little white box that they attached to a balloon and sent up into the atmosphere. the government told people that it involved gathering weather data. It used a very unusual type of pulse modulation. In most cases they used a CW (continuous wave) oscillator and pulsed the signal. This turned out to be a very efficient conversion of electrical energy to etheric energy. I very recently started collecting radiosondes. I never saw a receiver. I found out that they were designed up at Brookhaven National Labs. I started to talk to people at Brookhaven and ran into a retired gentleman who used to work there. He told me that the design was originally done by Wilhelm Reich. That piqued my interest.

The story goes that in about 1947 Wilhelm Reich handed the U.S. government a weather control device, a device that would do DOR-busting [DOR: Deadly Orgone Energy, a form of subtle life energy discovered by Reich]. Reich thought that if he could decrease the amount of DOR that storms would not be so violent. [DOR comes about through the contact of orgone, or life energy coming into contact with a radioactive source.] The government sent the device up there into a storm and it did reduce the intensity of the storm.

The government liked it, and they started another phase of Phoenix Project where they designed these radiosondes and started launching them in large numbers, maybe 200 to 500 per day. The radio in these things had a range of about 100 miles. If they used so many of them, one would think that receivers for them would be commonplace. I used to be a collector of radio receivers. I have over 100 in my personal collection. I have never seen a radiosonde receiver. I have heard of them but I have never seen them.

Q: How does this relate to what was going on with these other projects?

A: The government could not tell the public these were weather control devices. What we are seeing here is actually the genesis of what became the Montauk Project,

UFOS, TESLA, AND AREA 51 by Commander X

which was a combination of Wilhelm Reich's work and the Philadelphia Experiment.

There were two separate projects going on in Phoenix 1. You had the invisibility aspect and you had the development of Wilhelm Reich's weather control. Toward the end of the Phoenix Project, by using some of Wilhelm Reich's concepts and some of the transmission schemes used from the radiosonde project, they found that you could combine the two factors and use them for mind control. Government circles would have me say 'mood alteration,' but mind control is what these idiots were doing.

Q: That's what the Montauk Project was?

A: No. Phoenix 1. After political circles found out about it they wanted it shut down. The people that were running it went to the military and proposed that they could use it to 'influence the minds of the enemy.' The military loved the idea, and let them use the old Montauk Air Force Base. Among the equipment requested was an old SAGE radar unit, which was on the base. The base was shut down and everything was auctioned off. The group then moved in from the Brookhaven Labs. That began what we call the Phoenix 2.

They spent the first ten years, from about 1969 to about 1979, researching pure mind control. They started out by taking the output of the SAGE radar, modulating the special wave that Reich had shown them from the weather control process, and combined that with something noticed from the Philadelphia Experiment work.

Q: So Phoenix 2 started in 1969?

A: Yes, in the period from 1969 to 1971. Phoenix 1 went from 1948 to 1968. The first part of the mind control project was to take an individual and stand them about 250 feet away from the antenna. The SAGE radar had a peak pulse power of .5 MW. The antenna had a gain of 30db. That means an effective radiated power of at least a gigawatt. It was nominally a gigawatt. Can you imagine what that would do to people? I think it's amazing these people are still here. It does things like burn out brain functions, create neurological damage, scar lungs from heat, etc. They tried this with a number of people and there were few survivors.

UFOS, TESLA, AND AREA 51 by Commander X

Q: Where did they get the volunteers for this?

A: They were just grabbing indigent people off the street and throwing them in front of the radar beam. That's the sort of nonsense that the government loves to do.

Q: Who was in charge of the project at this time?

A: Dr. N. and P.

Q: Any particular agency?

A: I'm not sure what the agency was. Now, somebody got the brilliant idea to put the subjects directly in line with the gain horn of the antenna. Lo and behold they got their result without burning the people up. They found out that by varying the phase modulation and the frequency hopping and the pulsing of the multiple phases that they could have profound effects on a person's mind.

Q: How many people worked at this installation?

A: About 30.

Q: Who authorized the use of the base?

A: The Air Force and the Navy. It was a joint project. there were both Navy and Air Force personnel involved. We have copies of the orders for the Air Force personnel.

Q: What was the cover story for the base?

A: They had none. It was a derelict base. It was abandoned. It was turned over to the GSA as surplus around 1969 or 1970 when they shut down all the SAGE radar systems. It was a 'non-existent' operation. It was a perfect cover.

Q: Where did they get the funding?

A: it was totally private.

UFOS, TESLA, AND AREA 51 by Commander X

Q: Corporations?

A: It didn't originally come from corporations, although it did in later phases. The original money came from the Nazi government.

Q: This is Phoenix 1?

A: No, this is Phoenix 2 and 3. In 1944 there was an American troop train that went through a French railroad tunnel carrying $10 billion in Nazi gold which they had found. It was $10 billion at the 1944 price of $20 per ounce. The train was blown up in the tunnel. It killed 51 American soldiers. The gold turned up ten years later at Montauk. This has been verified. That money was used to finance the project for many years as the value of gold went up. They spent all of it and ran out of money. That's when they tapped on ITT, who funded it.

ITT was owned by Krupp in Germany. In terms of personnel, many of the civilians and scientists there were all ex-Nazis who came from Germany both before and after the war ended. The project was under U.S. government surveillance.

The intelligence community knew what was going on and the CIA monitored everything, as did other government intelligence agencies. The field of players who actually operated on the base was small, between 20 and 50. The funding was entirely private. After 1983, Senator Goldwater found out about it and started an investigation. He couldn't find any trace of government funding.

P. was the metaphysical director of the project. He was Air Force. After he left, Dr. U. took over. They had an electronics expert, Dr. Z., who came over from Germany in 1946 with Werner von Braun. Probably the reason that they ran out of money is that they had a total of 25 bases around the United States to support. The last of the bases shut down August 12, 1983. The base at Montauk, where all the stations got their zero-time reference from, shut down and the other two remaining bases went down with it.

Q: What about some of the mind effects?

A: From what I recall of the program, as I was part of it, I was subject to the mind control field not as I initially went into the program (because they wanted me initially for my sensitive abilities) but later. I was assigned to the indoctrination of the younger recruits. The first indoctrination turned out to be a disaster.

I told them I didn't want anything to do with the program, and they put m in front of this mind beam, and it did do damage to me. Finally, someone said, 'shut it off, he isn't going to give in to it.' And they shut it off. Others were affected much more seriously than I was. The effects were generally really bad. It could burn your brains out.

Q: Were any of the Montauk subjects given psychotropic drugs?

A: I don't think so... They used one drug which was used in connection with the Reich programming to make them more receptive. I don't remember the name of it.

Q: Did they use the mind-altering chair [that had been obtained by the government from the aliens]?

A: The prototype came from the aliens. Beyond that we are uncertain. This chair was essentially a mind amplifier. The government would have specially trained individuals sit in the chair and generate thoughtforms, which would be amplified and transmitted. They could transmit the signal and put people in a pre-orgasmic state where they would be receptive to programming.

It worked very well and they found other capabilities. They found that it could work in time. They had a psychically trained individual sit in the chair and generate a thoughtform of a vortex that connected 1947 and 1981. That's exactly what they go -- a time tunnel they could walk through. There was a series on television at one time that portrayed this concept fairly accurately. These were some of the earlier capabilities. They started going forward and backward in time. That was the last phase of the Phoenix Project.

Q: When did this time machine get going?

A: Around 1979 or 1980 it was fully operational. This transmitter had enough power to warp space and time. The individual in the chair would have to synthesize the vortex function because they didn't have the technical capability to do that. It can now be mechanically synthesized.

UFOS, TESLA, AND AREA 51 by Commander X

They did some other things. They had the subject in the chair think of some creature, and the creature would materialize. They had the individual in the chair think of all the animals at Montauk Point charging into town, and that's exactly what happened. They almost had the power to create a being. The problem they had was that what they created only stayed as long as the mind amplifier was on. The power was somewhere between gigawatts and terrawatts. Tremendous power. The vortex could have a diameter of about five miles.

Q: Can you describe what this looked like?

A: It's like looking into a peculiar spiral tunnel which was lit up down its entire length. You would start to walk into this thing and then suddenly you'd be pulled down it. You didn't walk through it as such. You were more or less propelled through it. You could go anywhere in space or time.

Q: Could you bring things back?

A: Yes.

Q: Have you ever brought anything back?

A: Yes.

Q: Could you continue your description of the tunnel?

A: Yes. The walls were solid but fluted. The tunnel was not straight but was sort of a corkscrew shape.

Q: If someone turned off the power, would you be stuck in the destination time and space?

A: Yes. You'd be stuck there. One of the first things they did was send recruits forward to around 6030 AD. It was always to the same point. Somewhere in an abandoned city where there was a statue of solid gold. When they came back they were asked what they saw. Whether they were expecting to find a different answer from person-to-person is unknown. They would look into the vortex and make sure the environment would support life before they sent people. They took samples.

UFOS, TESLA, AND AREA 51 by Commander X

Q: Are there potential futures that people could be sent to?

A: No. Once you make the connection with the future the line becomes fixed to that point.

Q: Can you change the present by sending someone to the past?

A: Yes. You can also change the present by sending someone into the future. Under certain conditions. The government is using existing time machines to go forward in the Montauk timeline.

Q: Are you saying that now the present can't be changed because we have established a time-loop through the future and the past?

A: Yes. That means whatever everyone is doing between the most extreme past point and the future they will be doing forever.

Q: What is the furthest anyone has traveled in the future?

A: 10,000 AD. It's a dreamlike reality. No one has picked up a tangible future beyond 2012 AD. There is a very abrupt wall there with nothing on the other side.

Q: Can you project yourself two hours into the future and meet yourself?

A: Yes, but it's very dangerous. The person who walked into the tunnel is out of phase with the person who comes out the other end. This did happen. The result is that the person just incinerates.

Q: Did they ever give you a weapon in case you ran into anything negative?

A: They didn't have to do that. The vortex could be arranged to follow the person, so that they could bring them back in if anything went wrong. They could see them on a viewer.

So they achieved a working time portal. At one point they had a situation where they had a 'monster from the ID' type creature come through and everyone went into a panic. They shut the transmitter off. The creature ate people and equipment. They had to go back and shut down the unit in Philadelphia in order to shut off the unit in the future so they could stop this creature in 1983. This was on August 12, 1983. The vortex locked on to the 12 August 1943 test and formed a loop.

All this occurred because someone planted the thought in the mind of the operator in the chair to generate this creature. It was an effort to sabotage the project. A lot of people thought the project had gone too far. When I worked for them between 1971 and 1983 I was so tired when I would get home from work. What they would do is that when my mission was over they would return me to a point milliseconds before I left. It would appear from one perspective that I never left. Of course, after I stopped working there, all that stopped.

If I may interject something here. There is a point about two-thirds of the way down the time tunnel where the person who is going through the tunnel perceives a large 'thump.' The person's consciousness leaves their body. There is a tendency to see things on a broader basis. I am sure there was some intelligence there. I would have trouble with the recognition of it. What they were trying to do at Montauk was to stabilize the perception process that would occur upon exteriorization from the body. They were trying to manifest that for some reason. We don't know what their purpose was.

Q: What was the role of the aliens in this whole thing, other than the prototype of the mental amplifier chair?

A: That would be another part of this that we haven't gotten into yet... Starting about August 6th, 1943 [After the Philadelphia Experiment], UFOs appeared over the Eldridge [Naval ship] for about six days. They were there during the [Philadelphia Experiment] test. One of the UFOs was sucked up into hyperspace with the Eldridge and it ended up in an underground facility in Montauk in 1983. It contained a charging device which some aliens made us go back and get for them, as they didn't want humans to have it. We don't know who they were. P. was concerned about an alien invasion.

Also, N. was called by the government to come and assist in the examination of a crashed UFO in 1947 at Aztec, New Mexico. Another crash occurred at Aztec about a year later. The first crash had greys on it and none survived. At least one occupant survived the second crash. The radar systems unintentionally brought down the craft. Radar was used intentionally after that until the aliens got wise to it. The occupant of the second crash was not a grey, and N. got to talk to it. N. asked it what the answer to the invisibility problems could be. He learned that he had to go back and do his homework in metaphysics. The nature of the problem was that the personnel on the ship were not locked o the zero-time reference of the ship. Humans are normally locked to the point of conception as a time reference, not a zero-time reference. The time stream lock allows the person to flow in synch with the system so interaction is possible.

Time locks are fragile. All of the power of the project disrupted the time-locks of the people on the deck of the ship [in the Philadelphia Experiment]. When the ship came back in time, the people didn't come back to the same reference.

N. realized that he needed a computer, as well as some knowledge of metaphysics in order to be able to lock the time reference of the people to the time reference of the ship. He built a computer in 1950 for the purpose. It was ready to be installed in 1952 and a test was performed in 1953 that was successful. They didn't go floating off into space when it was over. At this point, the Navy cancelled the project and changed the name to Project Phoenix.

A lot came out of the negative effects of the Rainbow Project. Some of it led to mind control research programs in the Phoenix Project. The invisibility research produced some Stealth technology as well as other highly classified projects.

In 1983, they decided to apply mind control to all participants in these projects in an effort to cover them up. They had also been working on another project: age regression. It was physical age regression. A person retained the memory they had from the older age and moved it forward in time. It took between 30 and 60 days for the body to complete the change to the new time reference.

Q: It seems like there are more people involved in secrecy than there are scientists. Where are all these people?

A: There are government agents and agencies everywhere that are concerned with keeping things secret. These days, the secrecy is applied more to the applications of hardware than the hardware itself. It's not like it was in the 1950's.

As an example, the guidance package for the new Minuteman X missile that was developed for the Air Force by Northrop is unclassified. There was no classification on the circuitry and the layout. It was so accurate that it could take a missile 5,000 miles and drop it down a chimney stack. The applications and what it was capable of was classified.

Q: Isn't a lot of this left out in the open to distract people from what is really going on?

A: Of course.

Q: What is the capacity of the gravity craft fleet of the United States?

A: I don't know. I know that they have built quite a number of these. When our astronauts first landed on the moon in 1969 they were greeted by a fleet of disks sitting on the rim of a crater. The astronauts asked their superiors if they knew about these disks. They were told 'Yes,' that they were American disks. The astronauts were angry at being used as public relations men by the government.

Q: Why spend so much money on the Stealth bomber when they have had this gravity technology for so many years?

A: Well, the aircraft combines two aspects for invisibility. One of the aspects relates to the construction and coating applied to the surface. The other aspect relates to an electronic type of invisibility package which is a result of work done on the Philadelphia Experiment years ago. Also, the Stealth has a secondary drive system which is very advanced and allows it to fly in space. The assistant director of NASA admitted that this came straight out of alien technology. He admitted this to the public.

Q: That's interesting in view of the government's apparent position with respect to covert technology.

A: There are breaks in the government secrecy programs that are starting to show up. More and more people are getting totally disgusted with government activities and attitudes and they are beginning to talk. Even MJ-12 in 1984 was about to break some information to the public about ET's and UFOs. They decided not to release it at that time. John Kennedy demanded that they release it within one year. He also demanded that the CIA get out of the drug business. They assassinated him.

Q: What are some of the other projects that relate to the capacity of factions operating within the United States Government and corporations to manipulate and control the population?

A: Well, between 1977 and 1978 a project called Dreamscan came on line. It ceased in 1979. The goal of the project was to gain the technical ability to enter into an individual's mind in the dream state and cause his death. there was a movie called Dreamscape which showed what they discovered they could do. The project was run by the Secret Government and managed by the NSA. The purpose of the project was to provide for a means of covert assassination. President Carter found out about it and had it stopped. The hardware is still intact and in storage. There have been attempts to put it back on line by various intelligence operations, some of which are said to involve [corporation name deleted].

Q: What else?

A: Around 1987, a project called Moonscan started. It lasted into 1989 and involved positioning mind control equipment on the moon for use on the population of the earth. It, like the others, has clear connections to negative alien activity.

Q: Who ran that one?

A: It was managed by an organization called [corporation name deleted], who have had other covert projects under their wing. At the time [deleted] was run by [deleted]. It is now managed by the Department of Defense as of 1988. there are three branches: Covert, Commercial, and Defense.

Q: Any other mind control programs that you can mention?

A: There was a project called Mindwreaker that would allow paralysis of the mind. The aliens were heavily involved with that project. It produced several neurological weapons, some of which are used on the B-1 bomber, which also contains a lot of alien technology. At the time, various alien species came and went out of [corporation name deleted].

There was one group called the K-Group, which was short for the Kondrashkin. They had pale skin that had a slight greenish tint and almost no hair. They looked like humans, and had to bleach their skin and wear wigs. They have been periodically involved with covert projects since the 1940's.

Q: Where has [deleted] been located?

A: In New York State, at Farmingdale, Deer Park, and Long Island. There were eight projects ongoing that also had to do with the development of weaponry against aliens. In 1989 the Orion group [of aliens] discovered this and destroyed the projects.

Q: What other research goes on at Long Island?

A: Research on scalar weaponry, like the one that destroyed the Challenger.

Q: I thought that the Challenger was destroyed by the Soviets using scalar weapons?

A: No. The Soviets didn't do that. The oddity with the incidents as far as the Soviets were concerned was that they pulled their ships about 150 miles out to sea before it happened. They were not the direct cause, which was a scalar weapon that they were trying to put into orbit and test. It accumulated a charge while the Challenger was going up through the atmosphere and turned itself on. That is what destroyed the Challenger. It might have been deliberate.

Q: What was the ultimate power behind the Phoenix projects and the mind control?

A: Ultimately, the whole thing is manipulated by the Orion [alien] group. The expectation was that they could use mind control to take over the populace in the 1990's -- no later than 1994 or 1995

They have also been doing genetic work in which they alter a human sperm and ovum to the extent that all offspring will produce hybrids with new characteristics.

UFOS, TESLA, AND AREA 51 by Commander X

Humans will mate and create children with alien genetics. That's one step beyond the average abduction scenario. There are other things happening with the human race.

Since 1947, there have been components of the 6th [alien] race incarnating on the planet. The 5th race was the Aryans. The 6th race humans are 100% telepathic -- the Secret Government and the Orion group sees them as a threat. They've been aware of it since 1948.

Q: What is the current situation with aliens?

A: Somewhat mixed and confused. There has been a lot going on all around the planet. In September and October 1990 there was an alien group from some other dimension that was attempting to invade the planet. They took down all the zero-time generators all over the country. The FAA was especially affected. The rogue group was stopped by another species. For many years, some factions of the Orion group depended on a ring of alien satellites that would sustain life functions. Those were wiped out in November 1990 by the same group.

Q: So there are positive light forces out there that are seeking to balance these negative activities by the Orion group?

A: Yes. I am not at liberty to tell you their identity.

THE PUERTO RICO PROJECT

The above interview offers the clues we need to understand another Top Secret project which has received little attention in the press. It was meant to be a hush-hush project, but word leaked out to UFO investigators around the world.

Late in 1984 two cargo ships crept into the port of Arroyo on the southern coast of Puerto Rico. Although they went about their business in straightforward manner in hopes of not attracting attention, the town was soon abuzz that the cargo carried by the Nautilus II and the Caribbean Adventurer were not simply the mainstays of daily life for the populace in Puerto Rico.

UFOS, TESLA, AND AREA 51 by Commander X

Inquiries were made among the port authorities, paperwork was checked, and thus another level of the cover-up was revealed. The cargo of the ships was allegedly "to be used by NASA." That much is true, but when a more in-depth inquiry the incredible truth surfaced.

One of the crew members of the Nautilus II spoke secretly to a reporter investigating the UFO flap in Puerto Rico, which included hundreds of sightings over the last several years. The crew member said that he had overheard an "official" who had travelled on board incognito saying that the cargo was nuclear weapons.

According to the crew member, the purpose of the nuclear weapons which had been unloaded were as defense against UFOs. Although this possibility was pooh-poohed by the press, there is a good deal of evidence that suggests that the crew member may have known what was going on.

The U.S. military, maintaining a number of bases in Puerto Rico, had determined that alien UFO bases had been established in the huge cavern systems on the island, and the military was quietly preparing defenses in case the so-far disengaged stance of the UFOs should change.

The caverns of Puerto Rico are one of the major cave systems of the world and were perfect, both for the concealment of alien and government bases. There are over 2,000 caves that have thus far been explored, and in addition there are huge deposits of vital minerals which make the area extremely valuable, possibly both for humans and for extraterrestrials. Is it possible that these supplies are being exploited by hostile forces from other planets?

Other factions of the press treated the reports of nuclear weapons seriously. The allegations of transportation of nuclear weapons into the country created a mini-flap in Puerto Rico, since the Treaty of Tecamachalco, which had been ratified in 1967, banned the deployment of nuclear weapons in Latin America. Puerto Rico had also been approved for membership in a Nuclear Free Zone.

Some Puerto Rican officials were highly incensed when the information of the nuclear weapons leaked out, and demanded an accounting of the United States. No accounting ever came, not even an acknowledgement of the concerns of these officials.

What was really going on in Puerto Rico? In years prior there had been an amazingly large number of UFO sightings, many of them engaged in incredible aerial antics and overflights. One of the most prevalent types was UFOs which would either plunge into the ocean around the island, or emerge from it, lending further credence to the possibility of concealed bases. There were other cases of individuals being followed and harassed by UFOs, and cases of individuals being disabled and abducted.

Above the town of Lajas in southern Puerto Rico, there had even been an aerial dogfight between a huge UFO and two American F-14 interceptor jets. The jets didn't stand a chance against the larger and more technologically advanced discoid craft, but before the eyes of stunned witnesses, were sucked into the giant UFO, never to emerge.

Was this really a skirmish in a covert war that was taking place in the skies above the Caribbean? It would not matter, for the military would never acknowledge the encounter. Other planes had engaged UFOs in the vicinity in the past, although the Navy naturally denied losing any of its planes.

One of the more stunning allegations that emerged about U.S. bases in Puerto Rico is that they were working on a Top Secret project called Excalibur. According to reports, Project Excalibur was a missile system that had been devised at Los Alamos, New Mexico, for the sole purpose of penetrating deep into the earth and destroying UFO bases! This project had already been mentioned in the reports of investigators, but was now confirmed. Project Excalibur has been known to researchers for a number of years, although little in the way of official documentation had been forthcoming.

Whether the existence of this Top Secret project was true or not, there have been many reports of explosions deep under the ground in Puerto Rico and in other areas. And there have been other amazing reports as well, including reports of buzzing and vibrations that are very similar to those which have taken place in the United States, such as the unexplained "Taos Hum" and other effects which may be the testing of electromagnetic weapons or some of the other technologies of Tesla.

Just what is going on in Puerto Rico? The information that we have received tells us that, among a variety of Top Secret projects, that deep within the cavern systems of Puerto Rico has been constructed an interdimensional door used as an entry point for time and dimensional shifting experiments by the U.S. government.

Strictly as a side effect of these experiments, we are told, are the "goatsuckers" or chupacabra which have plagued Puerto Rico in recent years. These were an unfortunate side effect of dimensional sweeps that were conducted by scientists working on the project, resulting in the release of these extra-dimensional creatures into the surrounding areas.

Who is running the Puerto Rican dimensional door? Although we have not received complete verification, we are told that it is a patriotic segment of the government who are interested in the defense of the Earth against hostile aliens. Let us hope that it is so.

UFOS, TESLA, AND AREA 51 by Commander X

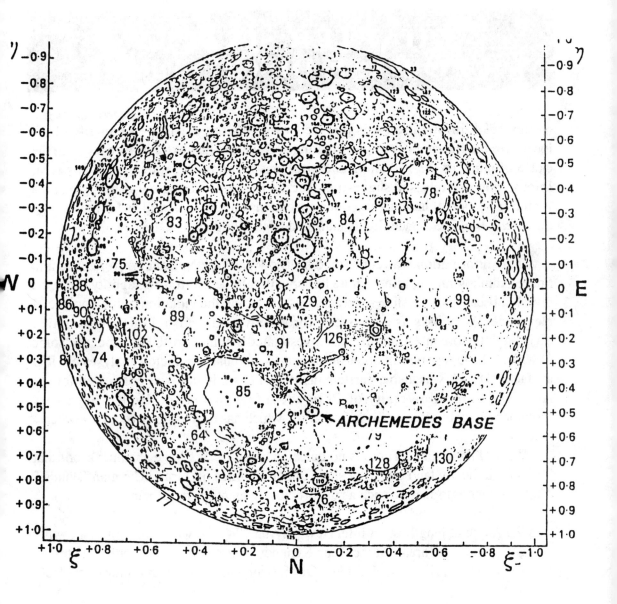

Alleged location of alien-human lunar base, portions of which are underground.

UFOS, TESLA, AND AREA 51 by Commander X

CHAPTER SIXTEEN
UFO BASE DISCOVERED

In 1994, information was released verifying the existence of an alien base located in the Bermuda Triangle, creating speculation that extraterrestrials were the cause of many of the 120 ships and airplanes that have disappeared in this area. Information was released suggesting that the Bermuda Triangle was a staging area for the collection of humans and other specimens on Earth, as well as for monitoring the activities of U.S. space flights launched from Florida.

This information was confirmed by Dr. W., of the National Aeronautics and Space Administration, although he has specifically asked for his identity not to be revealed because of possible repercussions, including the violation of his Top Secret clearance. W. works as a physicist for NASA, and leads a UFO investigative division in the organization. He has investigated a number of UFO incidents, some of which I have reported in other of my works, and has conducted analyses of UFO artifacts. Dr. W. was also a participant at a UFO study that was conducted for the Air Force at the University of Colorado a few years ago. He has been very forthcoming about offering his information, which has been little covered elsewhere.

According to him, "There is no question about it. We have absolute proof that UFOs periodically land on a tiny shoal about 50 miles off Grand Cayman Islands." This links into information such as the previously mentioned cases in Puerto Rico.

W. spoke at length in an interview, saying, "We've wondered for years what was going on in that area. At the Cape Kennedy control center we would get strange electronic blips during launch times. Our instruments showed that they came from this area.

"Then there were all the famous disappearances of airplanes and ships in the Triangle. On top of that we had several accounts of UFO sightings from some of the islands. So when two ships and the plane disappeared, we decided to have a look.

UFOS, TESLA, AND AREA 51 by Commander X

We took a small boat full of electronic equipment out to one of the small islands and sat and waited."

Two weeks after positioning themselves the research team, led by Dr. W., saw three bright circular UFOs descend toward the horizon. Although the objects were captured on film, the research group re-positioned to another island where they judged the UFOs to have landed, and in the following week they saw three objects, perhaps the same ones, land and then take off again.

Dr. W. noted that, "This time we were very close. We could see two small windows in the tiny craft. The UFOs themselves were shaped like tuna fish cans and could hover and fly sideways. Someday I suppose the full report will be made public, but not yet. All I can say is when we got very close to the objects they took off with incredible speed and headed into the sky in the direction of Bermuda."

Dr. W. also indicated a link between radio signals which were received at Cape Kennedy during launches. His conclusion?

"The Bahamas is perfect, because these [extraterrestrial] creatures can watch Cape Kennedy and electronically record everything that happens there. And they can also gather Earth specimens by capturing planes and small boats which venture into the Triangle. They can continue capturing the occasional ship and plane... and make it look like a strange natural phenomenon so people won't get frightened. People will just say, oh well, there goes another plane in the Bermuda Triangle.

"My own theory is that a large mother ship hovers several hundred miles above the Triangle and watches everything that goes on there. Then when there is going to be a moonshot or something of interest, it sends down the small reconnaissance UFOs.

"Those are the ones we saw on the island and, based on what they were doing, we can say that they were monitoring our... moonshot.

What Dr. W. did not speculate about are many ill-fated NASA launches which have taken place from Florida and nearby locations. Were the extraterrestrials responsible for these, sabotaging them by electromagnetic or other advanced weapons, or was some other factor connected to the Secret Government responsible?

UFOS, TESLA, AND AREA 51 by Commander X

At this point there is no way to confirm or deny these possibilities, but I would request that readers of my work who have information on this possibility, please forward it to my publisher so as to aid in building a database for research.

UFOS, TESLA, AREA 5l by Commander X

Human-looking ETs commonly referred to as "Nordic types" seem to be working hand-in-hand with the Grays and are part and parcel of a sinister plot by the New Work Order to rule Earth.

CHAPTER SEVENTEEN
NEW WORLD ORDER

Researching and writing books about the behind-the-scenes masters of world control as I do, a pattern starts to become obvious, one which has been with us for a very long time, even if it has been rarely recognized by the man in the street. That pattern, in my opinion, is what is most commonly referred to by the politicians and the media lackeys as the New World Order.

If you obtain all of your information from the government and Secret Government-owned media, then the term "New World Order" will more than likely offer you a warm, fuzzy assurance about a stable and wonderful future that will be experienced by you and your family -- something on the order of a combination of Tomorrowland and "It's a Small World" at Disneyland -- rather than cause you any undue fear. I don't see the New World Order quite that way.

I believe that there are many reasons for concern about the reality of this matter, and that we have not been told the whole truth about this "new order of the ages" as envisioned by occult secret societies of the past, or about the real plans which are coming to pass during these short days prior to the year 2000.

While world leaders are happy to toss the term "New World Order" about in their press releases and speeches, they are also quite careful to never really define the concept, or to place the idea into a historical context in which it can be understood. This is all in their favor, certainly, since if people really knew what the origins of the New World Order were, they might be moved to overthrow the plan entirely, and to tar and feather its supporters in the government and in the media.

I'm here to give you the other side of the story, the side they don't want you to hear. I'm here to give it to you straight.

UFOS, TESLA, AND AREA 51 by Commander X

NEW WORLD ORDER HISTORY

This touted New World Order is the ongoing plan of the Secret Government, specifically one small group within it called the Fabians, which were founded in London, England, in 1883. It is not within the purview of this book to outline in detail the ultimate beginnings of the Fabian philosophy and conspiracy -- all the way back to the occult Illuminati and to other secret conspiratorial forces which still exist today, and which provide the basis to their plans -- but the Fabians owe their philosophy to just these same secret sources.

Politically speaking, their plan is to proceed slowly but surely toward the total control of the world through the gradual infiltration of members of the Secret Government and their fellow travelers into key positions of control in the world. This control will take place in the political, financial, and other spheres of world control.

The Secret Government has gone about this plan in a very workmanlike fashion ever since the inception of their secret, conspiratorial group.

Regardless of what one believes about their philosophy, one must admire their industry and their dedication to the long term New World Order project, for the Secret Government has succeeded wildly in its schemes to overthrow existing systems and to take over the world. And, even more impressive for students of what they have accomplished, they have done it without even a whimper being raised by the American people whom it will mostly affect. That definitely took some planning on their part, and shows us that our enemy is a wily one and should not be underestimated in his ability.

While the Russian Soviet system may be forever blown to the wind (and may not be, depending on who one asks -- there are a number of political analysts who feel that glasnost is in fact another expression for "playing possum"), the Secret Government can be said to have succeeded famously in their plans for instituting world control.

UFOS, TESLA, AND AREA 51 by Commander X

Beginning with Great Britain where the Secret Government front group Fabian Society was first formed, the group took pains and took their time as needed, gradually infiltrating the policy making groups and governmental systems in England until that country became a cradle-to-the-grave Secret Government state in everything but its name. And so it remains until this day.

There, in England, toward the beginning of the 20th century, we first see the term "New World Order" being tossed about by elitist planners, men who were quite straightforward in their written works in admitting that the "common man" is simply clay to be molded in their hoped-for creation of a super world state.

There, among prominent members of the Fabians, we see the manifestos for turning the world into a system of total world control being written and published and read and hewed to as policy in the higher circles of government.

And yet, who is aware of the plans of the Secret Government for the world, besides a bunch of conspiracy-minded-crazies who are shouting that the sky is falling? Do any universities in any country teach that the Secret Government has gradually dyed the banners of the world Red? Is anyone moved to scream from the rooftops that our freedom has been taken away?

After gaining a secure foothold in Great Britain, the Secret Government quickly moved on to America. Working through well-financed subversive groups, academics, politicians and other idealogues were churned out and prepared for usefulness in achieving the One World superstate.

It is these same men who have since shaped education in America, rendering it into a cookie-cutter conveyer belt for the production of academic, political and worker drones. It is these men who have turned the newspapers, radio, and television in America into instruments for total social control and crowd management, rather than conveyors of information. They have instituted their form of gradualist control in a slow but sure transformation, until this country no longer bears much, if any, relation to the ideals upon which it was founded.

UFOS, TESLA, AND AREA 51 by Commander X

THREAT TO FREEDOM

Did someone say that the American Constitution is the force that is preventing a totalitarian takeover of this country? Give me a break, please. It is only too obvious that the document is only given lip service by the lackeys in Congress and in the presidency, and is set aside any time the ruling elite wishes to further consolidate their power.

The Bill of Rights? A document to be revered, honored, and respected, unless it gets in the way of police state tactics or the political ambitions of an up-and-coming politician.

Unfortunately, it seems that no one much cares what has happened to those documents guaranteeing the freedom of the people, either. People who view the compromises wrought upon the Constitution and the Bill of Rights as criminal are viewed as fools and fear mongers. Personally, I think a helping of deep fear is long overdue amongst the sheeple of this country.

There are many of the powerful elite in America who were groomed by Secret Government mentors with the overall plan being to further the implementation of the New World Order. There are numerous programs in the works that further this terrible program of social control, including plans for the replacement of American currency, the taking away of all guns from the people, the wiretapping of private conversations, and the institution of computerized identification cards, allowing for the tracking of the citizen from womb to tomb.

THE UNITED NATIONS: A NEW WORLD ORDER SCHEME

One of the most pivotal plans of this New World Order has been the institution of the United Nations, brought into being at the mid-point of this century, by Fabians and spies connected to the Secret Government. In an supposed effort to ease tensions between countries the world over through mediation, the U.N. is in fact a Trojan horse.

The U.N. was patterned after Woodrow Wilson's stillborn League of Nations toward the beginning of this century, and was instituted at the behest of master Secret Government elitist and Wilson puppetmaster

Colonel Edward House. This is the same man who wrote the novel Philip Dru: Administrator, which boldly exposes the plan for the Secret Government takeover of America and the world. Naturally the book is no longer in print.

On the surface the United Nations is intended to further the unification of the governments of the world into a single peaceful global community ("It takes a village"), in which all racial, ideological, and economic differences are pacified, mollified, and completely ironed out. It proposes to "teach the world to sing", like in the Coke commercial, and for all racial groups to be united in harmony (and preferably in picturesque ethnic garb).

But -- even if we will be allowed to sing and wear native dress when the New World Order comes to pass -- when have the controllers ever wished for the masses to have real democracy or power or any real say in their future? The answer is never. The answer is that they never will, even if supposed conservative Rush Limbaugh says so.

The real purpose of the United Nations is to emasculate the sovereignty and military power of individual countries, consolidating national laws, suppressing any dissent over the removal of human rights, outlawing "naive nationalism" amongst the nations of the world, and to control commerce and communications of all sorts.

The United Nations is meant to eventually replace the governments of the world with a single world government, much like the one represented in George Orwell's futuristic novel 1984. This is the age old utopian dream.

While a one world government promoting peace and harmony may sound like a great idea, especially if you are enamored by the crystal love and light ideas of the New Agers, I for one suspect the motivations of those who are furthering this plot. Why would the super-rich owners of this world have any interest whatsoever in giving away their ownership of the planet? I simply don't find the prospect very likely that that will happen.

WHAT IS HAPPENING TODAY?

In this book I argue that we may be at the point where the Secret Government-influenced controllers are ready to launch their coup de grace on social and political freedom in America, and to institute something akin to a Russian slave state in this country. They no doubt will use the high technological weapons that I have described in this book, and which have been provided by their own alien controllers. To one who is interested in freedom, it certainly seems that way lately.

It seems like every month some new governmental atrocity happens here in the States. Here in America, the iron fist of the government continues to drop suddenly upon unsuspecting and freedom-loving citizens, and nothing whatsoever in the way of protest is raised in the above-ground media.

If the erosion of Constitutional freedom in America are not enough to give one great concern about the future, then the consolidation of power, and the collaboration of technology and fascist, elitist control under the banner of the New World Order should.

Look at the plans, look at the programs, look at the experiments that government and intelligence agency scientists are currently engaged in, and decide whether the plans of these ultimate controllers are completely benign. Finally, look at the way that members of the Secret Government have compromised our future by their collaboration with the aliens, a group that certainly has no interest in the future of the human race.

A New World utopia? Then why the emphasis in intelligence agencies into such matters as technologically and chemically induced mind control, including brain implants intended to turn individuals into nothing short of robots? Why have the aliens been given free reign in their projects of harassment, experimentation, and mutilation?

Why the sudden interest in electromagnetic crowd dispersion and control? Why the concern for the precision electronic and computer monitoring of populations, including spy in the sky satellites with imaging capabilities so precise that they can read your newspaper over your shoulder while you sit on your patio? Why the interest in flying saucers that can be used to subdue any portions of the world who do not go along with the New World Order plan?

UFOS, TESLA, AND AREA 51 by Commander X

Are these Frankensteinian scientific methods the tools of men who are interested in providing freedom to the populace and creating a utopian future for the masses, or are these the tools of men who see the common person as a slightly more intelligent form of cattle, cattle who need to be controlled and who are nothing more than a resource for their own evil purposes.

"Live every day as if it were your last," is good advice. Not only because it is an impetus to appreciating the transitory moments of this life, but also because it may inform us of the gravity, the graveness of our situation at this time in the history of the Planet Earth, and encourage us to do something about it.

And when we see a continuing consolidation of control by the elite Secret Government conspirators and their alien masters, when we see new incursions upon our freedom with almost every passing day, that is enough warning, enough reason for us to open our eyes wide and decide that something must be done to change things.

For me, writing this book is part of that attempt to turn events around, and to prevent the plans of the controllers from reaching ultimate fruition. Time, I suspect, is short, although I desperately hope that I am wrong in this matter.

We must act quickly, with force and with accuracy, to combat these plans of the Secret Government and their alien controllers. We must work to protect the hard-won barricades that prevent free citizens from being enslaved to forces from outside of this world, and we must anticipate the next move of the controllers and do something to effectively counter it.

I suspect there will be many who view my assessment of the world situation and control (as well as the cosmic one) as insanely alarmist and terminally paranoid, and I suppose only time will tell who is correct in that respect.

It comes down to standing up for individual sovereignty and the rights of all persons, in any way that you can. It comes down to standing up for freedom in my country, in your country, in any country. It comes down to viewing oneself as partially responsible for what is taking place in the world, and to adopting a commitment to turn things around. It comes down to drawing a line in the sand.

I hope that this book will provide insights into what may be coming to pass in the future, and more than anything I hope it will encourage you to do something to prevent the alien darkness that is looming.

SECRET GOVERNMENT PLANS

Researcher Val Valerian, in his Matrix III, compiled a list of the plans of the Secret Government. I find that I agree with it in most instances, and it provides a ready checklist in attempting to understand the intentions of the men and women who are behind the New World Order. Here are its applicable points:

-- A One World Government and New World Order with a Unified Church and monetary system under their direction.

-- Destruction of all national identity and pride.

-- Destruction of all religions except that which they deem proper.

-- Mind control of every individual through what Brzezinski calls "technotronics", creating human-like robots and a system of terror.

-- An end to industrialization and production of nuclear power.

-- Legalization of all drugs and pornography.

-- Depopulation of all large cities into camp systems in the countryside.

-- Suppression of all scientific development not deemed useful.

-- Destruction of a significant amount of human population by limited wars, bacteriological warfare, chemical warfare, and electronic warfare. Three billion people must die by 2000. The Global 2000 Report produced by Cyrus Vance detailed this, and was accepted by Carter on behalf of the U.S. government. The U.S. population must be reduced by 100 million by 2050.

-- To weaken the moral fiber of the national and demoralize workers in the labor class by creating mass unemployment. Demoralized youth and workers will resort to alcohol and drugs. The youth will be encouraged to rebel, thus ensuring the destruction of the family unit.

-- To keep people from deciding their own destinies by means of one created crisis after another, then "managing" such crises. This will confuse and demoralize the population to the extent where massive apathy will result. FEMA is designed for the crisis management. States of apathy will be similarly induced through chemical means, such as inducements to consume fluorides and other harmful substances in the food and water, producing death and illness and insuring a continuing flow of capital into the medical establishment.

-- To introduce new cults and continue to boost those already functioning, which include various types of degenerative music.

-- To press for the spread of religious cults.

-- To export "religious liberation" ideas.

-- To cause a total collapse of the world economies and engender total political chaos.

-- To take control of all foreign and domestic policies of the United States.

-- To give full support to the supranational institutions such as the UN, IMF, BIS, World Court and cause the fading out of local institutions.

-- Penetrate and subvert all governments, and work from within them to destroy the sovereign integrity of nations represented by them.

-- Organize a world-wide terrorist apparatus and negotiate with terrorists whenever terrorist activities take place.

-- Take control of education in the United States with the intent and purpose of destroying it.

Ponder these points carefully.

IS IT TOO LATE ALREADY?

Fortunately, not all of the signs are grim. I have seen somewhat of a change take place in the last few years. In the area in which I live the odds are close to one out of ten that a person will have at least an idea of the owners of the planet, and perhaps will even have a very good idea of them and the tactics of control which they employ. People are also beginning to understand the alien threat, and to realize that this is a fact, and not just something from an episode of the X-Files.

The market in books that expose the Secret Government and alien control is very popular and continues to grow as the word is spread about who is running the show. And groups of persons interested in the restoration of human rights meet regularly in most towns in this country, and many of them actually get things accomplished, primarily in the area of education of people who do not understand what is going on. They are already providing a thorn in the side of the conspirators who are in power, particularly through the dissemination of information which renders the Secret Government and their plans vulnerable.

Worldwide, I notice that changes are also taking place. Around the world there is an uprising of sentiment which is bringing people to power who actually represent the governed rather than the rich governors and their alien masters. In Venezuela, in Thailand, in Mexico, in Italy, in Georgia, in the Ukraine, and elsewhere we see parties and candidates arising and triumphing against long-embedded party political machines. This is indeed a hopeful note.

No, not all is grim.

I hope that this book aids in your own understanding of the world situation and what can be expected to take place in the future. At the very worst, my assessment may cause people to examine more closely the freedoms they have, and to cause them to protect them with more care against possible incursions in the future. The threat to our freedoms is real, do not doubt that.

I count on your assistance in creating a better future for the people of this planet. I urge you to resist alien control.

A FORMER MILITARY INTELLIGENCE OPERATIVE'S EXPOSE AND SHOCKING PERSONAL BATTLE WITH THE "SECRET GOVERNMENT" AND ITS EVIL COLLABORATION WITH A GROUP OF ALIENS KNOWN AS THE "GRAYS"

In some areas bookstore owners have reported "serious threats" by sinister individuals, as well as calls made to their personal residence in the middle of the night, warning them to remove from their shelves all the writings of the author known as Commander X.

In several previous works, this retired military intelligence operative, through his close ties with members of the CIA and other "highly positioned" contacts, has dared to speak out about the on-going battle between a group of extraterrestrials and branches of the military who refuse to go along with a Top-Secret agreement made with the "Secret Government." This agreement, made in the 1950s, calls for an exchange of technology between the Grays and the "Secret Government" that operates with a clearance 15 degrees above that of our own President.

Now for the first time anywhere, Commander X reveals the intimate details of his own horrifying contacts with the Grays and his on-going harassment by certain branches of our military. In the newly released Commander X Files, the author dares to make the following disclosures.

8½ x 11 $39.95   ISBN-0-938294-32-6

✔ **ESCAPE TO ABDUCTION**
✔ **ALIEN-SECRET GOVERNMENT COLLABORATION**
✔ **THE SECRET WAR**   ✔ **TIME BOMB**
✔ **IN THE MATRIX**
✔ **NEW WORLD ORDER**

NEW

ADDITIONAL BEST SELLERS BY "COMMANDER X"

Order From:
INNER LIGHT PUBLICATIONS, BOX 753, NEW BRUNSWICK, NJ 08903. Add $4.00 to total order for shipping. Credit card holders may call 24 hours a day: 908 602-3407. Free catalog on request.

COSMIC PATRIOT FILES
From the "Committee of 12 to Save the Earth"
Edited by Commander X

Hundreds of topics are covered in this closely typeset, two volume—shrink-wrapped—set that has been specifically compiled for the student of New Age conspiracies.

Included are such intriguing, mind-boggling subjects as: • Founding of the Secret Order of the Illuminati. • Mystical significance of the layout of Washington, D.C., and the placement of the Washington Monument, the Pentagon and actual streets and other geographical locations. • Identification of global leaders involved in a conspiracy to formulate a "New World Order"—or "One World Government," and why they would want to do so. • Use of unmarked helicopters in keeping an eye on dissident citizens (including UFO witnesses). • Hitler's use of negative Occult influence and how these influences are still alive even today! • The responsibility for controlling the economy of the world. • The Great AIDS Cover-up revealed. • Activities of the "Brotherhood of Shadows" exposed. • Why Kennedy was shot.
ISBN: 0-938294-06-7 8-1/2x11—two volumes **$39.95**

2 VOLUME SET

THE ULTIMATE DECEPTION
by Commander X

This is a conspiracy that leads right up to the front gate of the White House, and involves the agreement forged between the military and a group of ETs referred to as the EBEs (short for Extraterrestrial Biological Entities). As part of this conspiracy, the government has literally "sold out" our citizens by extending to the EBEs the right to abduct humans and to plant monitoring devices in their brains, in exchange for technical and scientific data. "Our only hope for survival," says Commander X, "is a second group of benevolent ETs—most commonly referred to as the 'Nordic-types'—who believe in the universal law of 'non-interference.'"
ISBN: 0-938294-99-7 6x9 **$15.00**

UNDERGROUND ALIEN BASES
by Commander X

Aliens have established bases around the planet. An ancient tunnel system has existed on Earth since the time of Atlantis. Entrance ways can be found in many major cities. Some government and military officials have taken the side of the aliens. Here are bizarre stories about underground bases at Dulce, New Mexico; Groom Lake, Nevada; South and North Pole; Mt. Shasta, California, as well as in the Andes. Here also are the first-hand reports of individuals who have been abducted, and have survived genetic experiments in these locations.
ISBN: 0-938294-92-X 6x9 **$15.00**

NIKOLA TESLA—FREE ENERGY AND THE WHITE DOVE
Edited by Commander X

Here are Top Secret revelations concerning a newly-developed antigravity aircraft currently being tested inside Nevada's remote *Area 51*, as disclosed by a former military intelligence operative. This aircraft, which can fly three times higher and faster than any officially recognized plane or rocket, is based upon an invention of Nikola Tesla, one of the greatest "free thinkers" of all times, who arrived upon our cosmic shores in order to shape our technical and spiritual destiny. Tesla, the author reveals, came from another place to alert the world of impending danger (World Wars I and II), while at the same time offering "helpful solutions" to our problems and alternatives by which to greatly enhance our lives.

In addition to previously unpublished material on Tesla's Other-Worldly "roots," here also are full details of the ongoing work of such modern-day inventors as Otis T. Carr, Arthur H. Matthews and Howard Menger, who have perfected alternative methods of propulsion.
ISBN: 0-938294-82-2 6x9 **$15.00**

THE CONTROLLERS
Edited by Commander X

We are the property of an alien Intelligence! "Our" planet is a cosmic laboratory and we are but guinea pigs to those who have kept us prisoners on Earth. Humankind continues to face an all-out battle with those who have kept us as their slaves for centuries. Down through history, they have been known by different names: The Soulless Ones, The Elders, The Dero, The Grays, The Illuminati and The Counterfeit Race. Yet, very few know the real identity and purpose of *The Controllers*, a strange, parallel race that is metaphysically programmed to do evil and, according to authorities, has complete control of our education process, major philanthropic foundations, the banking system, the media, as well as dominant influence over all worldly governments.
ISBN: 0-938294-42-3 8-1/2x11—shrink-wrapped **$19.95**

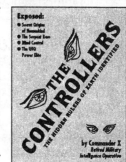

THE PHILADELPHIA EXPERIMENT CHRONICLES—EXPLORING THE STRANGE CASE OF ALFRED BIELEK & DR. M.K. JESSUP
Edited by Commander X

For the first time, a survivor tells his remarkable story of a career of brainwashing by the military, and how he eventually came to unlock the mysterious facts about what happened decades ago. Also reveals the case of famed scientist Dr. Morris K. Jessup, who died under mysterious circumstances because he knew too much about the Philadelphia Experiment.
ISBN: 0-938294-00-8 **$15.00**

THE SECRET DIARY OF COMM. ALVIN E. MOORE

On the evening of July 19, 1952, radar operators at the Washington National Airport spotted strange targets on their scopes. Soon they realized the National Capital had been invaded by a huge squadron of UFOs. The military ordered jets to intercept and clamped down a national hush-up. Even Capt. Edward Ruppelt, head of the Air Force's Project Bluebook, was not allowed to investigate and was rudely sent away when he arrived in Washington. When the news leaked to the press, the government explained the incident as a "temperature inversion."

Early the next morning, a caretaker on the estate of Commander Alvin E. Moore, near Herndon, Virginia, noticed something very peculiar - branches of trees had been damaged, and then he spied a gaping hole directly beneath the damaged limbs, which exuded a sulfurous odor. He ran to get his employer, who had not yet left for the Technical Information Branch of the Bureau of Aeronautics in Washington, where he was division head.

The caretaker and Commander Moore carefully exhumed a still-warm and obviously manufactured object. The old-line Naval officer soon realized the thing could not be a meteorite, and was obviously manufactured -- for it had three machined sides and a fourth that appeared to have broken off. As he carried the object back to his house, his brow pinched in puzzlement, for his years of technical experience in aeronautics told him that this was not from an aircraft -- indeed, he had the impression that the thing was from out of this world!

Commander Moore had the object analyzed and discovered that it was not of natural origin.

So began the Commander's involvement with the strange world of interplanetary craft and government coverups and harassment by the various agencies that employed him. For Commander Alvin E. Moore was NO crackpot. Previous to the fall of the object that was to change his life, he had chalked up an impressive career. Educated at the U.S. Naval Academy, the American University, George Washington Law School, etc., he had nearly enough credits for a Ph.D. As a line officer of the U.S. Navy, he had specialized in aeronautical engineering and was patent engineer and attorney for the von Braun team of space scientists at Huntsville, Alabama. He has also been a CIA Intelligence operative, and while with the Agency became impressed with UFOs, finding it somewhat strange that the CIA also seemed to be fascinated with them.

THE SECRET DIARY OF COMMANDER ALVIN E. MOORE tells the inside story of the UFO that crashed outside Moore's home and how the wreckage was stolen from his office safe in a government building. Here are his most private thoughts recorded before his passing as to what the U.S. really knows about the purpose and origin of flying saucers and their occupants. Dare you learn what he discovered???

Many of the things he FOUND OUT were utterly astonishing. For example, Moore believed that UFOs exist in mysterious "sky islands" much closer to Earth than many of us believe. The Government, he said, suppressed this secret to prevent mass panic, mainly because some of the more negative "Skymen" cause fires, explosions, aircraft crashes, train, auto and ship wrecks, diseases, wounds, burns and killings.

Moore also explains the UFO abduction mystery, why some contactees are given misleading stories, and how the Men in Black play an important part in silencing UFO witnesses.

Commander Moore's findings are so shocking that they were never widely distributed during his lifetime and are only now available on a limited basis for those wishing a private glimpse into what the U.S. military and the Secret Government really knowns about "space visitors."

TO ORDER: Send $18.95 plus $4 for Priority Mail to: GLOBAL COMMUNICATIONS, P.O. BOX 753, NEW BRUNSWICK, NJ 08903
Credit card holders may call 24 hours a day 908 602-3407.